ATC & Weather
Mastering the Systems

Second Edition

Also by Richard L. Collins

Flying Safely
Flying IFR
Flying the Weather Map
Tips to Fly By
Thunderstorms and Airplanes
Instrument Flying Refresher (with Patrick Bradley)
Flight Level Flying
Air Crashes
The Perfect Flight
Confident Flying: A Pilot Upgrade (with Patrick Bradley)

ATC & Weather
Mastering the Systems

Richard L. Collins

Second Edition

**Understanding how to work
air traffic control and weather
to best advantage**

Aviation Supplies & Academics, Inc.
Newcastle, Washington

ATC & Weather: Mastering the Systems
Second Edition
by Richard L. Collins

Aviation Supplies & Academics, Inc.
7005 132nd Place SE
Newcastle, Washington 98059-3153

Published 2001 by Aviation Supplies & Academics, Inc.
First edition, *Mastering the Systems: Air Traffic Control and Weather,* published 1991 by Macmillan.

Printed in the United States of America

05   04   03   02   01   9   8   7   6   5   4   3   2   1

ASA-ATC-WX
ISBN 1-56027-424-7

Library of Congress Cataloging-in-Publication Data:

Collins, Richard L., 1933-
    Mastering the systems: air traffic control and weather / Richard L. Collins
        p.   cm.
    "An Eleanor Friede book."
    Includes index.
    ISBN 0-02-527245-4
    1. Meteorology in aeronautics.   2. Air traffic control.
    I. Title.
TL556.C583 1991   90-47391   CIP

For Rowland Bedell. Friend, aviator, healer—*Gone West 4/4/90.*

Contents

Preface

to the Second Edition

The air itself hasn't changed much since I started flying about 50 years ago. The weather now is very much like it was then, and I would be hesitant to say that weather forecasts, especially for more than a few hours in the future, are much better. The basics of aerodynamics are the same, and airplanes are shaped very much like the ones that were flying when I began.

Most of the similarities end there. In 1951 we flew virtually without avionics except in some of the more advanced general aviation airplanes. There was little general aviation IFR flown, and our primary contact with air traffic control was with the few towers that existed. With no VHF, we communicated on low frequency. The other day when tuning a nondirectional radio beacon on frequency 278, I thought that a familiar number. It was. All towers transmitted on 278. We'd listen on that and transmit on 3105 kilocycles, as they were called at the time. A fairly substantial number of operations at airports with control towers were by radioless airplanes, controlled by light gun signals from the tower. Rocking the wings would acknowledge that you saw the signal. After I got my instructor certificate I spent many hours teaching flying in Taylorcrafts with no radios at Adams Field in Little Rock, Arkansas. We used the pavement some, but when the infield was in good shape we'd conduct most operations on the grass. Many airports at that time maintained grass runways along with pavement because the tires lasted longer when you landed on grass, and that surface was more tolerant of a little drift at touchdown when there was a crosswind.

Instrument flying caught my eye early. My father, Leighton Collins, had long been a proponent of light airplane instrument flying, so I took to it naturally. He sold me his IFR-equipped Piper Pacer for a nepotistic sum; I got my rating in it in 1955 and started learning about the inside of clouds and the air traffic control system.

There was no traffic control radar, and the FAA kept up with IFR airplanes based on position reports. You talked to what is now the flight service station, and the person there would relay your progress to the center. You had to hit your estimate over a station within three minutes, which was no mean feat on the first leg when you had no idea of what the actual wind aloft might be.

They were easy times, flying was fun and relaxed, and we were all having a ball with our airplanes. Then in a blinding flash, on June 30, 1956, everything changed. I was standing in front of a barracks at Fort Rucker where I was stationed. It was one of those hot Alabama days, with haze hanging heavy in the still air. There were several of us there, standing in the shade—no cool place to go because there was no air conditioning—when a buddy came over and told us that two airliners were missing out west somewhere. We tuned up a radio and learned that the airplanes were missing in the vicinity of the Grand Canyon. A TWA Constellation and a United DC-7 were presumed down. There was no suggestion of a collision at first, though it would have seemed a possibility.

That evening we did what soldiers usually do, and the next morning we were up early, wondering about the two airliners. Later that day it was announced that they had been found. No survivors. An apparent collision. I told you about those estimates on which they tracked airplanes. Both these flights had estimated a point known as the Painted Desert line of position at 10:31. At 10:31, the day of the accident, controllers heard a garbled message which was later interpreted as "Salt Lake, United 718…ah…we're going in." The identical estimates over Painted Desert were tragically prophetic. Because one of the aircraft was VFR and one IFR, the controllers didn't try to separate them.

The accident has often been called the forerunner of today's air traffic control system. It caused a flurry of regulation and opened the funding floodgates for radar. VFR flying was prohibited above 18,000 feet. (These two airplanes had been flying at 21,500 feet.) Within a few years an outline was in place for the air traffic control system to become about what it is now. In the '60s, there were a number of collisions between general aviation and airline aircraft. These hastened the implementation of terminal control areas (what is now Class B airspace) and other regulated airspace that was envisioned in the wake of the Grand Canyon collision.

It is within this air traffic control system that we fly today. Although it came about in an evolutionary way, it bears no similarity to the system that existed when I started flying. Everything has changed. It is not simple, nor is it as complicated as a lot of the other things we learn to do. Some of the easy nature of flying is gone, but some of it remains—probably more of the nature of the '50s remains in flying than in other activities. It counts for a lot that people are still flying to be able to do the things they want to do. We got a nice illustration of this in the summer of 1990.

Our house is on one of the routes flown by Marine One, the presidential helicopter, between the White House and Camp David. We have seen the helicopter go over many times, usually with one or more others in attendance. One day I was out in the yard and heard the distinctive sound coming. Marine One was lower and slower than usual and came across the adjacent golf course, flew right over our house, started a climb, and disappeared over the horizon. A few weeks later it came back to nearby Frederick Airport, where President Bush (the first) and party boarded cars and drove out to Holly Hills Country Club for 18 holes of golf. The President had seen the course many times in flying over and thought it looked like a good place to play. The day I saw them fly over lower, they apparently got the closer look that clinched the deal. He later landed on the golf course for his outings.

There is still a lot of old-fashioned freedom in flying. Go where you want to go, when you want to go. That is something we need to protect. One thing we must do is encourage others to take it up so that it might resume growth following a slump that spanned the '90s. Another important part of the process is to be a cautious and well-informed user of airplanes, to make individual contributions to a continuous improvement in the safety record. Honing flying skills is an enjoyable challenge. So is a continuous study of the relationship between the two systems in which we fly, the air traffic control system and the weather system at hand for the next flight. By thinking of them together you can have a satisfying relationship with both.

Richard L. Collins
Ijamsville, Maryland

Chapter 1
The Myths

Aviation, as much as any other activity, is beset by myths—things people will to be true even though there is no basis for them. I will always remember an old codger looking through the window of my Cessna Cardinal RG, seeing one of the first Stormscopes residing in the panel, and saying, "That will get you in trouble, boy." The implication being that more information is bad, that it can only lead you down the primrose path. The information has to be correct, even precise, but the simple fact is that the more information we have, the better. Flight itself might be a pure thing, driven by the controls of the airplane and the relationship between man and machine, but we live in a modern world. The airplanes we fly have to reflect this if we are to take full advantage of what is out there. There is still a place for the VFR stick and rudder pilot who marvels in the myths of flight past, and the basic skills can't be neglected, but there is so much more opportunity now. Anyone who doesn't immerse himself in the full array of what is available is truly missing out—just as a pilot who doesn't master the basic skills is missing out. And one of the keys to full participation is found in trashing the myths—the things that some generations of pilots have willed to be true.

Ocean of Air
One myth that has always been fun is about how the airplane flies in an ocean of air and only the ground track and ground speed are affected by wind. When I was working full-time for *AOPA Pilot* we printed a story in which a pilot related an increased rate of climb as he ascended into an increasing headwind. Letters, we got letters. "Don't you dumb bunnies

know that an airplane in flight is unaffected by wind?" That is one of the older myths, one that has cost lives over the years.

The air, like the ocean to which it is compared, ebbs and flows. One of the most important things we learn as pilots is that this is true. Flying a high-performance jet we don't see it much when going fast, but it can have a profound effect when we are flying slowly, as on approach or just after takeoff. Flying slower airplanes we can see it in wider phases of flight. I have flown my Cessna P210 for more than 8,000 hours. Because it is pressurized I fly it in the 19,000 to 21,000 (Flight Level 190 to FL210) area often. And the air up there can be strange indeed.

Flying it at FL190 from Little Rock, Arkansas, to Maryland one winter day, with a tailwind of over 100 knots, I don't think I ever saw what could be called a stable indicated airspeed. There was a jet stream core above, with winds well in excess of 100 knots. All the air carrier crews were complaining mightily about turbulence above FL210, and many were slumming with me below FL200. The turbulence wasn't bad, light at most, but there was an almost continuous jiggle and the indicated airspeed would range from 120 to 150. That was a result of changes in wind at my level, as the effect of the even stronger wind above undulated.

How it Works

It is true that an airplane in flight is basically unaffected by a steady wind. Changing wind does have an effect, however, and wind almost always changes with altitude and often with distance. On the flight back from Little Rock just related, I was storming along with that tailwind, and when over Elkins, West Virginia, I started thinking about the descent. There would be two strong factors working: One was a high ground speed; the other would be a decreasing tailwind on the descent. Coming down is more difficult with a decreasing tailwind. Why? If the airplane is moving across the ground at 270 knots, as mine was that day, with a true airspeed of 170 knots, and is descending to a level where, with less tailwind, it will have a ground speed of 200 knots, the airplane actually has to decelerate by 70 knots. That is almost heresy, saying that in this sense the airplane is ground referenced, but this is how it works. If it is going to go 70 knots slower across the ground, the simple fact is that it has to decelerate. That day, with the power slowly reduced to be kind to the engine, and with the power as low (20 inches of manifold pressure) as I like, the airplane would descend only at 500 feet per minute with the airspeed at the top of the green arc. Normally it will descend at 800 to 900 feet per minute.

Interface

In a situation like this the interface between the air and the airspace becomes a definite factor. Having flown in from this direction many times, I knew the controller would come at me with a clearance to cross either 20 or 30 miles southwest of Martinsburg at 9,000 feet, so I would have 10,000 feet to lose. Usually I figure five miles per 1,000 feet for a normal descent, but with the ground speed high and with the decreasing tailwind I figured I would need a lot more distance. When I was 100 miles from Martinsburg, I explained the predicament to the controller. At first he couldn't comprehend that I would want to start down so far out, but he came through with a clearance to cross 20 from Martinsburg at 9,000. And it took every inch of the distance to nurse the airplane down.

Other Places

You can see the effect of changing wind on the airplane in a lot of other places. The notorious downburst airliner accidents are clear evidence. It has happened dramatically both on approach, to the Eastern 727 at JFK, the Delta L-1011 at DFW, and the US Airways DC-9 at Charlotte; and on departure, to the Pan Am 727 at New Orleans. A rapidly decreasing headwind or increasing tailwind means the aircraft has to accelerate just to maintain airspeed; in both cases the combination of the downdraft from a storm and the decreasing headwind or increasing tailwind can exceed the ability of the airplane to accelerate.

Not Necessarily Wild

In some references to downburst and the effects of thunderstorms, pretty wild numbers are used. I have seen a downdraft of 5,000 to 6,000 feet per minute mentioned. Perhaps this is possible on a one in a million basis; it sure doesn't take that much to get the best of an airplane, though, and the Eastern 727 accident at JFK, which was widely studied, is a perfect example.

The aircraft was on approach, following other aircraft whose crews had reported hazardous wind conditions. The crew heard the report from one of the other aircraft; there was quite obviously a thunderstorm on the final approach course, yet they continued. The 727 penetrated the storm when it was between 500 and 600 feet above the ground, on final.

From the NTSB report: "The increase in headwind of about 15 knots and possibly an updraft produced a reduction in the rate of descent and the airplane moved slightly above the glidepath as it descended between

600 feet and 500 feet. When the flight descended through 500 feet, about 8,000 feet from the runway threshold, the airplane was passing into the most severe part of the storm. The vertical draft changed to a downdraft of about 16 fps (960 feet per minute) and the headwind diminished about five knots. As the airplane descended through 400 feet, the downdraft velocity increased to about 21 fps (1,260 feet per minute) and the airplane began to descend rapidly below the glide slope. Almost simultaneously, the change in the direction of the horizontal outflow produced a 15 knot decrease in the airplane's headwind component, which caused the airplane to lose more lift and to pitch nose down. Consequently, the descent rate increased."

The report went on to state that preceding aircraft encountered similar but perhaps less severe conditions and that one captain needed near-maximum thrust to keep his aircraft from losing altitude, was not sure of his aircraft's missed approach capability, and felt compelled to continue to a landing.

In this case the downdraft strength was nowhere near that quoted when trying to scare folks about downdrafts, and the wind change was a relatively mild total of 20 knots. Yet it was labeled a "very strong thunderstorm," and it bested this crew.

Bigger Not Better

Another significant myth was addressed in the report on the accident at JFK. When discussing the effect on other aircraft making the approach, the report stated "…the pilot of N240V, a Beechcraft Baron, was able to limit the altitude loss caused by the wind condition with less difficulty because of the different flight characteristics of the smaller aircraft and because he was flying at a higher-than-normal approach speed." The 727 was at a higher speed, too, but the significance of this is that light airplanes do better at adjusting to wind changes because, in the approach configuration, they are operating at a lower relative power setting than a heavy jet and have better acceleration characteristics. The reason a 727 can approach as slowly as it does is because of all those high-lift devices that unfold from the front and rear of the wing. You don't make lift without creating drag. The pilot of a Baron or other light airplane uses quite a low percentage of power to track a glide slope; a 727 pilot uses a substantial percentage of power while flying the approach because of the increase in drag.

Normal Climb

Back to normal operations, when climbing it is possible to learn something about the winds aloft. For example, when you are climbing in low-level turbulence and then reach smooth air with a headwind, there is usually a spike in the rate of climb because the wind velocity increases when you reach the smooth air. Lower, the velocity is lessened slightly simply because the air is disturbed. Likewise, when climbing with a tailwind, a decrease in rate of climb likely means an increase in the velocity of the tailwind.

One final item on how the movement of air affects performance is settling, called subsidence, and rising. This might be found in and around cumulus clouds where the air tends to rise in the clouds and settle in the clear areas around them. Basically, air flows around and into a low-pressure area and out of and away from a high-pressure area. That really tells you all you need to know about rising and settling air in that case. In convective outlooks, the National Weather Service sometimes refers to UVV when discussing a developing low—which stands for upward vertical velocity. Simply put, air settles in a high and rises in a low. If ever you fly a trip and feel that the true airspeed wasn't quite up to par, chances are you were in the middle of a high. One other place where we see subsidence on a fairly large scale is around thunderstorms. All the air that goes vertical to build the storm has to come from somewhere, and when flying near a storm you can see some substantial airspeed losses caused by subsidence.

So changes in airflow do affect an airplane in flight. Perhaps the easiest place to see this is on a gusty day, on approach. The airspeed jumps around like mad because of changes in wind direction and velocity.

Multimotors

After World War II, many of the myths that developed came from former military pilots. The postwar FAA was populated by them, as was the civilian pilot corps. But we apparently got more bomber and transport pilots than fighter pilots. One reasoning was that fighter pilots, when they stepped out of their '51 for the last time, were not further attracted to flying. Having survived that, why press on? The result was that much of the policy, procedure, and custom was shaped by those who, in the military, had referred to smaller airplanes as "peashooters." This has had a lasting effect on aviation.

For years people who flew singles IFR were viewed by many as crazy. The old codger related earlier, who thought a Stormscope in a single could only lead to trouble, probably firmly believed that to be true. The fact that FSS weather briefers ask you what kind of airplane you are flying is another throwback, as if the airplane and not the pilot determines capability. For years, insurance companies would give you a lower rate on hull insurance on a multiengine airplane based mainly on what they thought to be true. Then they did their sums, found they were losing a lot more money on twins than singles, and reversed their field.

The myth here is really not related to the single. You can ask anyone on the street what happens if the engine on a single quits and he can give you a correct answer: The airplane comes down. The myth is with the twin, because a layman would probably say that the airplane flies on one engine to a safe landing on an airport. The record over the years has shown this to be only partly true. In many different studies, including one by the NTSB, it has been shown that your chances of being terminated because of the failure of an engine are four times as high in a light twin as in a single. And the single record includes all the homebuilts. There's nothing intrinsically bad about homebuilts, but neither their engines nor their fuel systems have to meet the standards of certified airplanes.

The actual numbers on IFR are interesting, too. Singles fly about a million and a half hours a year in instrument meteorological conditions where piston twins fly about a quarter of a million hours less. In one study covering three years, it was found that nine twins crashed with fatal results because of a mechanical failure of an engine as opposed to one single. This suggests the singles are a lot safer. But the singles had, in the period, twice as many fatal crashes related to systems failures so, in total, the accident rate related to failures is about the same. On singles, though, you can now address this question with redundant systems. Standby vacuum to run the instruments is readily available for all airplanes and this has proven to be the most critical system. Fortunately, with the advent of standby vacuum systems and more pilot-awareness of the subject, the number of singles lost after a failure of the vacuum system has declined.

The Fit, Please
Where does the business about singles and twins fit into the overall scheme? It is pretty simple, and it also relates to the Baron/727 analogy cited on page 4. Until something happens to the machine, its relationship to ATC and weather systems is related more to the way in which it is operated than to the number of engines or the weight of the airplane.

Thunderstorms, for example, have proven lethal to all sizes of airplanes, up to and including 747s, in all phases of flight. And thunderstorms are no threat to any size airplane unless the pilot of the airplane chooses to fly into one of them beasts. It's that simple.

There is a Difference

Having said all that, you might think a myth is being perpetuated about a single being safer than a twin, but this is far from the truth. Each is as safe as its pilot, who has to make a judgment on how and where the airplane is operated. The airplanes are equal until and unless an engine quits, and although the actual mechanical failure of an engine is always a possibility, it is relatively low on the list of factors that lead to fatal accidents. How the airplane is operated relates directly to equipment, too. If the airplane is operated with the bare minimum of equipment, then there are places it should not go. Thunderstorm areas are a good example. The air traffic control system can offer some thunderstorm guidance, but you can't count on it. At times they close off sections of the country and reroute flights because of the presence of storms. Some controllers will tell you what they see in the way of weather on their scope. Some will even offer vectors around what they see. It is pure myth, though, to imagine that the air traffic control system is, or ever will be, a primary weather avoidance system. That ability comes from information on the panel and from what is seen; responsibility resides in the left seat of the airplane.

Legal and Safe

A definite myth exists regarding the Federal Aviation Regulations, the rules by which we fly. The myth is that if it is legal, it is also safe. There is no truth to that. It is fact that to disregard the rules is dangerous, but it is equally dangerous to feel that abiding by the rules is all it takes to minimize the risks in flying.

Again, thunderstorms are a good example. There is no rule against flying in thunderstorms, but it is surely not a wise thing to do. One reason there is no rule against it is that it is all but impossible to write a rule about an element that is so variable. The FAA proved its lack of atmospheric understanding with a proposed rule about icing in 1989. The rule would have prohibited flight in known icing conditions and would have defined "known" ice as any time the temperature was below 5°C and visible moisture was present. That would have pretty well prohibited flying on other than clear days all winter. The proposal was dropped and one would surely hope the rulewriter was embarrassed over the idea.

The VFR weather minimums are often quite unsafe for VFR flying, as is proven every year. This is especially true in mountainous terrain. For years the FAA did not differentiate between daylight and dark in VFR weather minimums, and the evidence on this lies in a fatal accident rate at night that is from three to five times as high as by day, depending on how you massage the numbers. Most accidents involving a pilot who continues VFR into adverse weather conditions, day or night, starting with good intentions, and a pilot who feels he is within the regulations right up to the last minute.

The rule on fuel says we have to land with only 30 minutes of fuel for a day VFR flight. Anyone who routinely plans this will eventually run out of gas. On many airplanes, we fly with somewhat less than the maximum fuel even after telling the line crew to fill the tanks and checking that the level is at least near the top. If the wings of the aircraft are not level, the shortage can be accentuated. On a Cessna 210, for example, you can easily be 8 to 10 gallons short unless you put special effort into packing that fuel in with the wings level. So how do you tell if the wings are level before pumping the gas? Simple. The ball in the turn coordinator or turn and bank will be in the center. The point is, for the person who would fly based on a 30-minute reserve, on the average fill you are probably 30 minutes down to begin.

Fixin'

Another area in which to beware of accepting the rules as any assurance of anything is in maintenance. When we don't operate for hire, the only requirement is for an annual inspection. That is as true for a simple single as for a turbocharged and pressurized single. Where it might be adequate for the smaller airplane, when they become complex they need to be looked at more often. And they need to be looked at thoroughly. From time to time I get calls from people with airplanes like mine, and one question they often ask is, "How much did you pay for your last annual?" It is usually three or four thousand bucks and I cringe when someone replies, "You got taken; I got mine for a thousand." What he really got was an inadequate inspection. Tales abound on airplanes that got "cheap" maintenance for a while and then cost tens of thousands of dollars to bring back into a true airworthy state. Especially as airplanes age, they need more expensive maintenance — the older they get the more we have to spend to keep them going.

People who decry the state of maintenance today and suggest new rules are right about maintenance being poor in some cases, but the suggested cures are Draconian at times. What we need is better enforcement of the current maintenance rules. The poorly maintained airplanes are the ones that come from shops where the major piece of equipment is the fountain pen used to sign off the annual. I got a call from a person who operates a fleet of Beech 99s and is in the market for these airplanes all the time. Some of the tales he told about the condition of some airplanes in the fleet were amazing. Many have just not been maintained to any standard.

Those are just a few examples of things that are legal but quite unsafe. As pilots we have to view the rules as a minimum framework in which to build a series of margins above and beyond the rules. Airline rules have the margins built in, but the fact that they change on a fairly frequent basis is proof enough that you can't write rules to cover every eventuality.

One final word: The words "FAA approved" are often used to try to cast a favorable light on something. Before biting on that, remember that virtually every airplane destroyed in a crash was "FAA approved" and every pilot was certified by the FAA. Although everybody should get the necessary approvals and certificates, they are not magic and guarantee little.

No Bad Guys

One myth that is always bothersome relates to the feds who run the aviation system—the FAA inspectors and the air traffic controllers as opposed to FAA management and policy folks. These operational people are cast in a bad light by some pilots, especially when the FAA launches an aggressive enforcement campaign. It is not their fault. If there is unfairness, the responsibility for it is usually rooted in the General Counsel's office and ultimately the Administrator's office. No controller or inspector is going to adopt a "take no prisoners" attitude without being made to do so from above. In fact, if they can do so without being discovered, many controllers and inspectors will cover for pilots on minor items, especially if they are satisfied there was no intent to err. On balance the controllers especially are truly there to help out. A pilot who yells at a controller because he doesn't like a routing is talking to the wrong person. The air traffic control system is rooted in procedures that controllers must follow, and none of them intentionally and personally send a pilot out of the way. The same thing is true of the airline pilots with whom we share the system. Some of them don't like little airplanes, but some of them probably don't like their mothers either. You can't expect everyone to love everyone else. But as long as we hold up our end of the bargain and operate within the system properly, and with a degree of professionalism, few airline pilots have trouble with general aviation operators.

PIREPs

All of us have bemoaned the paucity of pilot reports on tops and icing and turbulence when we know that there are airline, commuter, and air taxi aircraft out there 24 hours a day. When you listen to some pilots discuss this, it almost sounds that a flood of pilot reports would forever scare away all our weather problems. Flight service station specialists always ask us to file pilot reports and seem genuinely appreciative when we give one.

A pilot report, though, is valid only for a short time and if it is re-ported by a pilot flying a similar airplane. Weather changes far too quickly for us to bet anything other than a wooden nickel on the conditions someone encountered even a few minutes before in a changing situa-tion, such as an area of developing thunderstorms. Ice is another one because there is a surface temperature rise on the airframe as indicated airspeed increases. So a jet might report no ice where a Bonanza is really getting loaded up. (The temperature rise is between 3° and 4°C at 11,000 feet and 140 knots indicated; at 320 it is about 17° at 11,000 feet.)

Even pilot reports of turbulence have to be used with care. Convec-tive and mechanical turbulence affect lighter airplanes more than heavier airplanes. But wind shear turbulence may seem worse in a heavier air-plane. Also, many types of turbulence change quickly with distance. If a cold front roars through, there might be moderate turbulence in the frontal zone over, say, Little Rock. Thirty minutes later it might be light there and coming up on moderate over Memphis, where it was reported as smooth 30 minutes ago. Pilot reports are great and should be encour-aged but, again, no magic.

Long of Tooth

When relating an accident that happened on his airport, a friend ex-pressed amazement that two pilots, each with over 15,000 hours, had flown their jet into a thunderstorm on an approach. The result was a total disaster. I tried to explain to my friend that having a lot of hours guarantees nothing. It is a pure myth to put total faith into a lot of flying hours because, in truth, only the next one counts.

It is true that experience is valuable, but in some ways experience can also hurt. If it breeds complacency, it is a bad deal.

Perhaps thunderstorms are one of the best examples on this. The airline folks will look you right in the eye and say that they do not fly in thunderstorms. But then you can go to a busy airport that is being raked by thunderstorms and, lo and behold, the airlines will be conducting near-normal operations. We have all seen it happen and, when the subject is raised I have had airline pilots ask if I could imagine the chaos that would result if they suspended operations because of storms. You can certainly see their point.

The Delta L-1011 accident at DFW was a prime example. One of the pilots remarked that there was lightning ahead. There was also talk of getting the airplane washed and of being in rain. The experienced pilots thus made a conscious decision to fly through a thunderstorm, as did the Eastern 727 crew at JFK. The Pan Am crew at New Orleans took off in one. In each one of the cases, the pilots had probably taken off or landed with thunderstorms around numerous times. Their experience told them it would work, their peers were out flying, and so they kept on going. Perhaps having many hours doesn't ensure that a pilot realizes that although a lot of thunderstorms are manageable in heavy airplanes, a few are not, and there is really no way to tell them apart. Experience is wonderful as long as suspicion and caution remain. You'd always rather have that combination. But if it comes to choosing between a 15,000-hour pilot who thinks he has seen it all, and a 1,000-hour pilot who is suspicious of everything and very cautious, I'd take the 1,000-hour guy.

Bucks

A popular myth is: Flying has gotten so much more expensive that I had to quit. In relation to what? Flying has actually gotten less expensive in relation to some things, such as housing in major metropolitan areas. Many other things have gone up relatively more than flying: restaurant meals, hotel rooms, new suits, and silk neckties on which to spill soup, to name a few. If you adjust the numbers for inflation, learning to fly in a Cessna Skyhawk today costs the same as it cost to learn to fly in a J-3 in 1951—and the new airplane is much more capable and comfortable.

Flying has always been expensive, and some of the things we buy today to make the airplane work better for us in the weather and air traffic control systems add expense. So does the extra performance we buy. But it still comes at a relatively low price. When I look at the combination of turbocharging and pressurization, and all the capability I have in the panel of my P210, I actually wonder how you get all that for so little in relation to what it cost to fly, say, a Bonanza when I started in 1951. Bonanzas then, with no IFR capability and with a cruise of about 140 knots and a practical ceiling of 12,500 feet, cost $25 per hour. Now I have radar, Stormscope, deicing, dual systems, 23,000-feet and 190-knot capability for less relative cost at $175 per hour, to say nothing of things like comfort and range and all the other gadgets. There is no question that you can sink $100,000 into the finest array of electronics for a single. That's an option, based on how much information you want to have as you fly in the system. But you can also fly plenty of IFR with a couple of nav/com/GPS radios and a Mode C transponder. Much of the expense today is at the choice of the user. For some reason, though, many people who fly are willing to grant inflation to everything but flying.

The question is not really expense, or cost; it is perceived value. Back in the good old days our piston airplanes flew as fast as airliners — which meant that you could get there more quickly in your own because you avoided stops and changes. The airlines didn't have ridiculous low fares on some routes, as they do now. And there wasn't the competition for disposable income in the heyday of general aviation. Today you can spend those extra bucks on a lot more things than you could 20 or 30 years ago, and most of the gadgets you buy today — boats and motor homes, for example — are more user friendly than airplanes.

No Myth

One thing is not a myth: Despite all the tribulations, the general aviation airplane remains one of the finest personal or business possessions. The freedom it affords is unlike anything else. The automobile comes closest, but the airplane is so much faster. You can do things with it that are otherwise impossible. One winter day, four of us piled into my P210 and flew to White Plains, New York, to have lunch with some folks there. We didn't leave particularly early, we had a good visit and a nice flight up and back, and we were home by four. As we rolled the airplane back into its T-hangar, we all agreed that nothing, nothing beats an airplane.

And all you have to do to make it work is learn to deal with the weather and air traffic control systems.

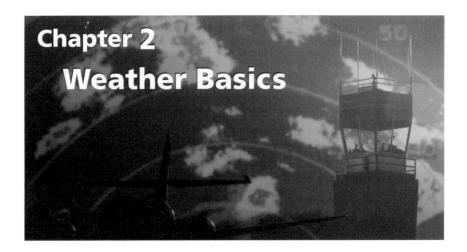

Chapter 2
Weather Basics

Some pilots wind their way through a flying career with only a cursory understanding of weather. They can probably read a sequence report (now called METARs) and they know that the wind shifts with a front. This knowledge of only the most basic of the basics seems to work, but it is a shame to stop here because weather is fascinating and learning as much as possible about it is fun. The more you know, the better you will do when the forecasts all go sour because you will be aware of the factors that lead forecasters down a primrose path. You'll know the conditions that suggest weather worse than forecast. And you'll know the telltales of really severe weather. Meteorology as taught to meteorologists is of necessity quite complex. But meteorology as it needs to be understood by pilots is much simpler and can actually be reduced to the interpretation of what you see and feel. The airplane is an excellent weather sensor and the fact that you are strapped to it means you have a ringside seat for the greatest show on (or above) earth.

Two of the more important items used in anticipating weather are the wind and temperatures aloft. The wind tells us about circulation, the temperature tells us about stability. (If the temperature drops more than 2°C per 1,000 feet of altitude, the air is conditionally unstable.) The pressurized Cessna 210 I fly is an especially interesting airplane from a weather standpoint because it operates in the middle levels, the high teens, and the low 20s, and what goes on at these levels has a profound effect on surface weather.

Number One

Throw out a challenge: What is the most basic, most important thing for a pilot to understand about weather? It has to relate to accepting conditions at face value. What you see is what you get. What is the most important pilot reaction to weather? Curiosity or suspicion. It is necessary to keep asking the question: Is that weather out there really doing what I thought it was going to do? By continually asking that question you lead yourself to a better understanding of weather because you will try to determine why the weather took an unexpected turn.

And weather can have its moments. Especially in the fall, winter, and spring the weather can be both violent and changeable. Forecasting the dynamics of weather is based on using a computer model, or models, integrating all current factors with history to develop an idea of what might happen. Once a storm system forms and starts moving it becomes fairly predictable. It's making a judgment of when the storm will form and how it will track that is tricky.

As I'm writing this, at eight on a January morning in Maryland, there's a nice sunrise out there though clouds are rolling in. The night before, on the news after the conclusion of the Super Bowl, a winter storm for the area was predicted. Sleet was to begin sometime after midnight, perhaps mixed with snow. The frozen stuff did not come because the low-pressure system did not develop and move as the computer thought it would. I have been watching the weather because I am supposed to go to Texas tomorrow and we need good weather there for video photography. The systems just haven't been strong and the forecasts have not been particularly good. So when the long-range forecast tells of a nice week in Texas, especially in the winter, there has to be some suspicion. Checking the weather for more than the next little while is fine if you want to do it, but making or changing plans based on a long-range forecast is an approximate activity at best.

Well Founded, Too

I have returned from Texas, and I am here to tell you that the suspicion of Texas forecasts was well founded. The trip to Grand Prairie, Texas, from Maryland was in good weather, but with a strong headwind. That evening the weather report on the local TV in Dallas was of a chance of some cloudiness the next afternoon with perhaps a shower. Improving conditions would follow. It still didn't look too bad to the computer.

Up early the next morning to fly, with nice sunlight but with an ominous cloud shield to the south. By noon it was raining big drops. By three in the afternoon there was thunder and lightning that persisted through the night and into the next morning. This was all laid to an "upper-level disturbance," and the official forecast was changed to one calling for scuzzy weather for several days.

There would be no more photography, so I thought I would go home. The forecast for the area was for low IFR conditions—300-foot ceilings and a mile visibility with continuing thunder and lightning—until a frontal passage in the afternoon. But as day broke, the thunder abated, the visibility was good, and we could see the jetliners on approach to DFW, just north of Grand Prairie, flying by at about 4,000 feet. The forecast of low IFR was incorrect. It is when the forecast is wrong that your personal weather knowledge will either make you home free or lead you down a primrose path.

The optimist in me said that the upper-level disturbance had moved to the north. That coupled with the radar picture on the morning news suggested that flying toward the southeast for a while would get me out of the area of disturbed weather. Then I could turn toward the northeast and go home.

Missed Clue

At the airport it appeared fine to take off. I looked at a clue—the wind-sock—and knew what it meant, but failed to go through a complete analysis of everything it meant. The wind was out of the west, favoring the north a little. Even the most elementary understanding of meteorology leads you to know that a front had passed. Whether it was the front that was forecast to pass later in the day, or a front that developed on a new surface low that was aided and abetted by the disturbance aloft was beside the point. Where earlier the wind had been strong and gusty from the southeast, it was now from the west. The primrose path down which I flew to the southeast took me through the front. There was no lightning—the Stormscope was clear—and the radar showed only green and yellow in the area where I penetrated the front. Yet the wind shear turbulence met at least one of the qualifications of "severe." There was no way to maintain altitude; the aircraft ascended from 3,000 to 3,800 feet with the landing gear down and the power almost at idle while the airspeed was kept as close to maneuvering speed as possible. The airplane was not largely bothered in roll, which it would have been in convective activity. It just had a bad case of the ups and the downs and it kept me quite busy for about five minutes.

How Better?
There was nothing hazardous about flying through that front at 3,000 feet. The level of comfort just wasn't as high as it might have been. What I did wrong was fly through a front using the best procedure for flying in an area where the weather is dominated by an upper-level disturbance. By flying low, I got the maximum effect out of the cold front at the surface. Other aircraft in the area, at higher altitudes, reported much less turbulence. Most of the time in Texas it would have been better to fly low, but not that time. By not taking the time to completely analyze a cross section of the airspace through which I would be flying, I went at a less than optimum altitude. Oh well, the rest of the trip home was okay, though it was almost all in cloud at Flight Level 190 where the forecasts had suggested I would be in the clear about halfway into the trip. The tailwind wasn't all that good flying home either. It was a different sort of system — it seemed to languish in Texas rather than move to the northeast — and the National Weather Service consistently missed forecasts for the next four or five days. They kept calling for rain in Maryland and we kept enjoying spectacular sunrises and sunsets.

What You See
When flying, what you see is about the best information you can get. On a January trip from Maryland to Little Rock, Arkansas, good conditions were forecast. All Maryland was to have were snow flurries. There was no turbulence called for on the area forecast. Ho hum. A piece of cake.

Not long after departure from Carroll County Airport in Maryland, the plot and the clouds thickened. There was strong overrunning, and the overrunning air was quite unstable. How can you tell? From the look of the clouds. The bottoms of them were dark; they had that old ripped look in some areas and circular sculptures on the bottom of the dark clouds in other areas. Lots of churning going on in there. I hadn't flown halfway across West Virginia when there was talk on the radio of developing areas of precipitation and there was soon what appeared to be a wall of snow to the south of my course. The churning clouds were based at about 8,000 feet so I tried to stay below, at 6,000 feet. But it was turbulent there; in fact, pilots at all levels up into the 30s were complaining of turbulence. It was clear to me that if I was to get where I was going, there would be some turbulent instrument flying. The back side of the weather was about 50 miles east of Lexington, Kentucky. I flew into the snow just southwest of Clarksburg, West Virginia, so I had about 150 miles of snow flying to do.

When I passed north of Charleston, West Virginia, the controller there said the snowflakes were as big as he had ever seen. On a day when it was forecast to be partly cloudy, they got nine inches of snow. Back home in Maryland they got four inches. To give you a little cross section of the atmosphere, I flew to Little Rock nonstop at 6,000 feet with a block-to-block ground speed of 143 knots, reflecting a 15-knot headwind component. The next day I flew back to Maryland at Flight Level 190 (19,000 feet) at an average block-to-block of 229 knots, a 50-knot gain. Actually on the trip back the ground speed was as high as 275. The air was full of jiggles all the way back. Although conditions in the lower levels were fairly calm both days, there was strong overrunning at higher altitudes and the day before there was enough moisture to create a surprise.

Picture

So what went on to cause snow that day? Whenever you see those churning overrunning clouds, it is a likely indication of a trough of low pressure aloft, with unstable air overrunning more stable air near the ground. There's also a moisture supply—witness the general dark color of the overrunning clouds. The fact that the air was generally turbulent at all levels was indicative of wind shear—greater velocity with altitude in this case. So this wetter and stronger than expected trough moved through rather quickly and dumped more snow than anyone thought it would. Flurries turned into an almost full-blown snowstorm.

The business about an upper-level disturbance or a trough aloft can illustrate an important point about weather. Years ago they didn't talk much about conditions aloft but now even the basic TV weatherfolk rattle on about the effect of the jet stream and about upper-level support for developing storm systems. Although the surface weather map does just what it says in showing you the locations of the systems on the surface, conditions aloft have everything to do with the way the surface map develops.

Low/Front

An oversimplification can be used to help understand lows and fronts and how they develop and move. First, think of one front that goes all the way around the Northern Hemisphere. In spots it is a cold front, in other areas it is a warm stationary or occluded front. Lows develop along the frontal zone, changing the nature of it as the circulation develops around the low. With a fully developed low, you might have the classic frontal model, with a cold front trailing southwest from the low and a warm front to the east. It is in the development of low-pressure systems that the upper-level support and jet stream business come in. The key upper level is at about 18,000 feet, or, the 500-millibar level, as shown on the 500-millibar chart. This is key to both the development and movement of surface lows. The support they talk about comes from colder air aloft creating instability and from low pressure aloft allowing air from the surface to swirl inward and upward, enabling the development of a complete circulation around an area of low pressure at the surface. Surface lows tend to develop when a trough of low pressure aloft moves overhead—a trough being where there is a southerly pooch in the winds. The tip of a trough is where a northwesterly wind aloft turns and becomes a southwesterly wind aloft. When a surface low forms it tracks about with the wind at the 18,000-foot level. If the wind is from 230° on the east side of the trough aloft, the low will track toward 050°.

As the low moves northeastward there's the classic cold and warm fronts with a southerly circulation east of the cold and south of the warm front, a southeasterly circulation north of the warm front, and an easterly flow north of the low coming around to a northwesterly flow behind the cold front. The farther away the low gets, the weaker the circulation. Finally the cold front stops moving, becomes a stationary front, and waits patiently for the next batch of upper-level support to come along and help another low get its start in life. You have perhaps heard the weather forecaster speak of a "return cold front." This is actually a warm front. How it comes about is from a cold front that moves through and becomes stationary to the south; then a new low forms on the front to the west and a brand-new warm front is created out of the clouds and moisture of the old cold front.

Waves

At times a stationary front will result in a lot of inclement weather, some of it relatively nasty, that suggests things are becoming active and a low is forming. This usually relates to low-pressure waves, sometimes called just plain waves, on the front. This phenomenon occurs when there is some instability aloft, but not enough to support the development of a full-fledged circulation around a surface low. The inward and upward swirl begins, creating an area of showery and often turbulent air along and north of the stationary front, but it never becomes a complete low. Instead, the wave moves up the front for a while and then dissipates. If you are flying north of a stationary front and fly from wet and bumpy into more stable conditions, and then into more wet and bumpy one or two hundred miles on down the road, the likelihood is that you are flying through a succession of low-pressure waves. These are more prevalent in the southeastern and south-central United States in the wintertime, where wintertime cold fronts often become stationary fronts. The first time you fly through a succession of waves can be somewhat perplexing. Our pilot mindset is that when we fly through some weather and then get on the back side of it, the battle is won. But with waves, the areas of poor weather recur until you fly out of the frontal zone.

Ground and Air Clues

How do we use this part of the grand plan to help us understand systems as they develop?

Just understanding circulation and temperature and moisture supply is a big help.

Your location and the direction of the wind is a big clue. Take that stormy day in Texas that was discussed on pages 14 to 16. When the wind was out of the southeast it was moving warm and moist air up underneath colder air aloft—that upper-level disturbance they were talking about. When the wind shifted to the west it brought cooler and drier air in at the surface; this resulted in an increase in stability and an improvement in the surface weather. But there was still instability and moisture aloft. That cold front at the surface was likely just a temporary incursion caused by the development of a surface circulation that had not been anticipated in the low-IFR forecast for all day.

When a strong cold front goes through, the wind generally shifts to somewhere between west and northwest. The closer to north, the more likely the front is to become stationary not too long after it passes through. If the surface wind becomes northeasterly soon after a frontal passage, that generally means the front has stopped. The circulation around the low is, in your location, too weak to sustain movement of the front. On the other hand, if a cold front goes through and you get one of those fine ripping northwest winds and cold air, you know that front is going to boogie on to the southeast for a while. To better visualize this, think in terms of the wind moving the front. If it is perpendicular to the front, the front moves. If the wind becomes more nearly parallel with the front, movement slows or stops.

Warm One

Warm fronts don't give signals that are as clear. Generally, if you are north of a warm front the wind will be from a southeasterly direction; when the surface warm front passes, the wind will shift to a more southerly direction and the weather will improve from the wetness and rain that is characteristic of the area north of a warm front. Often, there will be what seems a warm front to the east of a low but no front will be shown on a chart. This is because all the conditions that define a front are not present. There has to be a wind shift, a pressure jump, a temperature change, and a dewpoint change for a front to earn its place on a map.

Warm Storm

Warm fronts can cause widespread areas of low ceilings and generally preclude VFR flying. For the IFR pilot, conditions are often okay, though there are always those embedded thunderstorm forecasts to make life interesting. Warm frontal storms are different than other storms in that the instability that supports them is above the slope of the warm front. The action is at a higher altitude and the thunderstorm sometimes doesn't influence the air beneath the slope of a warm front too much. Warm fronts have rather shallow slopes, though. With an east-west warm front, starting at the surface position of the front and heading north, the slope might average 1:100. That means that 100 miles north of the front, the slope would be just over 5,000 feet above the ground. When pilots who flew IFR in times before weather radar and Stormscopes want to fly as low as possible in such conditions, that is the reason. They are trying to stay below the slope of the warm front.

Frozen

Another weather treat you have to beware of when flying north of a warm front is freezing rain. This condition doesn't occur all that often, but when it does it can be the ultimate bad scene even if your aircraft is equipped for flight in icing conditions.

When rain falling out of warm air above the frontal slope falls into below-freezing air beneath the slope, presto, freezing rain. It is in this condition that the old saw about climbing above ice came to be born.

And that might well work—at least for a while. But unless your airplane is gifted with an unlimited supply of fuel, you may not be able to keep it in that warm air forever. Any approach to an airport that is reporting freezing rain is chancy at best because the icing condition will get worse as you descend, the landing might have to be made with an opaque windshield, and a go-around might be impossible. At the first hint of freezing rain, the best deal may be to go to the nearest airport where it is not occurring and land. Rules of thumb about weather are always suspect but generally, with an east-west frontal orientation you'd get the biggest potential change in conditions by flying either north or south—which way depending on where you started, the capability of your airplane, and the results of a careful check of surface weather.

Lots of Lows
At times we see a weather map that is not so clear-cut. One might show several low-pressure systems scattered around in half of the country with stationary fronts or even no fronts. Usually the weather is a little inclement over a wide area. A map like this may not be clear-cut, but it does give a clear message. No storm system was well organized when the map was drawn. Stand by. Low pressure aloft will come along and aid and abet the strong generation of one of those lows. Then that low will become a predominant feature on the map, with well-defined fronts, and will take off cross-country.

Troughs
When a cold front is not quite a full-fledged cold front, it is called a trough of low pressure. Any front is actually an area of low pressure, with the lowest pressure registered at about the surface position of the front. A trough is a similar protrusion of the isobars away from a surface low. The prudent pilot, when told of a trough, might just as well think of it as a front. It may become one later; even if it doesn't, enough characteristics of a front will be there to suggest that it be treated as a front.

R-rated: Violence
Where we have to really excel in our understanding of weather is in relation to the violent events in the sky. This is generally related to severe thunderstorms.

Many pilots contend there are so many wolf calls about severe storms that they hardly know when to suspect that storms might actually be a factor on a flight. And certainly they are forecast more often than they actually occur. When the forecaster prepares what is called a convective outlook, there is often nothing going on. What he is saying is that in an area (usually shown in red on the morning TV shows) the combination of ingredients required for severe storm formation might come together later in the day. It will be watched like a hawk and specifics will be covered in convective SIGMETs if storms do develop.

The jet stream—more accurately a jet core which is an area of faster moving air within the jet stream, advancing in the general circulation—is key to the formation of most severe weather. A jet core isn't likely to cause a lot of trouble as it heads toward the southeast, but when it reaches the tip of the trough aloft and heads northeast trouble can start. Colder air over warmer air is one thing and a jet core ups the ante. The air rotates within a jet core, and there is descending air in the right front quadrant. This descending air is heated as it is compressed toward the surface, which further increases surface and low-altitude temperatures beneath the cold air aloft. It also pushes air, resulting in convergence about 100 miles to the southeast of the position of a surface cold front. It is here that spectacular storms develop.

Wind at FL180

A wise weather sage once put this into context for me. He said to always look at the winds at the 18,000-foot level, even if flying lower. If they are strong at that level, and out of the south or southwest, and it is warm at the surface, watch out for fireworks.

I was thinking of that one day when headed for Savannah, Georgia, from New Jersey. The headwind was humongous below 10,000 feet and it seemed that I was hardly making progress. Higher, up in the 18,000-foot level, the wind was even stronger, and at every level it was much stronger and a bit more southerly than forecast. My wise weather sage identified that as another bad sign. More southerly means more moisture; stronger means the developing low is stronger than anticipated. That morning I had drawn a low on the map, in northern Mississippi, with the classic frontal model around the low. The forecasts for along the way were benign, but I wrote one ominous note on the briefing form: "occ to W ATL." The briefer had mentioned an occlusion to the west of Atlanta and an occlusion—where the circulation around a low is so strong that the cold front catches up to the warm front—is usually indicative of a wild ride.

Because of the strong wind, an unscheduled fuel stop was made at Raleigh–Durham, North Carolina. There the conditions were, well, eerie. The surface wind wasn't particularly strong, but it was quite warm, and the pressure was dropping like a stone. "But," the briefer said, "all the activity is to the west of your route."

Not far south of Raleigh, we had the activity in sight as well as on the radar and Stormscope. Just to look at it was fearful. The monster had a yellow-greenish appearance, and although we never got closer than 25 miles to it, there was weird turbulence; at one point I had an overwhelming desire to put more distance between my little green airplane and the big green storm. When I told the controller I needed to turn 45 degrees to the left he approved it, but added that I was well clear of the storm. I just wasn't as well clear as I wanted to be.

When we got to Savannah the wind was howling out of the south. Where the forecast had called for a 15-knot southerly wind, it was over 30. An airliner reported a quite strong wind at 2,000 feet and a significant wind shear. The news that night carried the rest of the story—big tornadoes in North Carolina. Although I heard no mention of severe weather in any of the briefings I got, the conditions were clearly there, and I felt both forewarned and forearmed through the use of simple basics about weather.

Occlusion

I suppose there can be what one would call relatively benign occlusions, but in studying the relationship of weather and airplanes, flying and watching, I have become quite wary of occlusions. There are documented cases of them creating severe turbulence many miles away from precipitation, and occlusions seem to have been a factor in the development of several severe weather systems that have been studied. A low doesn't move very fast as it is developing, or deepening as it's called, and the circulation increases as this happens. The occlusion is more likely to occur in connection with a strong developing low that gets wrapped up quickly, the fronts form, and boom, the cold front catches up with the warm front. The low, or cyclone as it is sometimes called, reaches its maximum intensity at about the time the occlusion occurs, and it often appears that the time of maximum disturbance is at and a few hours after the cold front overtakes the warm front.

Other Basics

There will be a lot more on weather in other chapters, but in thinking of basics some things are useful to keep in mind. You may read them and decide the author is paranoid about weather, but the truth is that there are tales and lessons learned behind each thought.

One wag offered the meteorological rule of thumb that clouds always form or dissipate at sunrise. The implication is that the weather reports made at night are not as accurate (nor do they come from as many places) as those made in the daytime. If you are an early riser, the forecasts are likely rather old when you get them. The pilot reports are few and far between. Like all of us do, the weather system (not the weather itself) takes a rest at night. What that tells the early-morning pilot is to move with an extra degree of weather caution. It may be clear, with stars in view, but if the temperature and dew point are close, fog may form at about sunrise. If it is a go, after you leave and after things start stirring, a bit of extra enroute weather checking is in order.

Always take a forecast as an educated guess. And never get upset because a forecast is wrong. It happens all the time. If the weather is worse than forecast, we had best build protection against that eventuality into our flight plan. If it is better than forecast, all to the good. Many times with a busted forecast you can study the situation and see that there were signs that it would go sour. Keys are in the comparison between the forecast and the actual surface weather, and between the actual wind and temperature aloft and that forecast. If the actual wind is substantially different from the forecast wind, that means the model they had of the atmosphere was incorrect, so all forecasts should be suspect.

When fog forms it then becomes a forecaster's job to predict when the fog will lift. That is not easy because other factors, perhaps unknown to the forecaster, have a strong influence on this. For example, if there is a bit of higher cloud cover, keeping the sunlight off the low cloud tops, the fog might take longer to dissipate. So it isn't a good idea to bet it all on a forecast of fog to lift by 10 A.M.

Finally, and again, the most basic requirement is the need for understanding. If we maintain an active curiosity about what is going on out there and approach it in a cautious and conservative manner, weather can be a well-understood friend as opposed to a capricious menace.

Chapter 3
Flight Planning and Weather

When I started flying—and let's not call them the good old days—most weather briefings were one-on-one from what is now called a flight service specialist. In some areas you could actually call the Weather Bureau (now the National Weather Service) and get your briefing from a real-live meteorologist. The meteorologists would go into great detail on the diagrams of temperature, pressure, and humidity aloft—looking for warnings of instability and resulting thunderstorms. But there was no weather or air traffic radar at that time and the only real information we got on thunderstorms was from station reports. (A thunderstorm is considered to exist whenever thunder can be heard by the observer.) We did not see many weather maps unless the briefing was the result of a visit to the facility. TV was not widespread in the early '50s and weather reporting in the media was far from the slick celebrity sport that it has become today. About the best you could do for a map, really, was to look at a day-old one in some of the newspapers. The only other available aviation weather information came from the scheduled broadcasts on low-frequency-range stations, at 15 and 45 after the hour, that gave the surface observation for the airport involved as well as for ten or twelve others in the area. These gave another clue: If there was static we knew there were thunderstorms around. As VOR stations came along, the scheduled broadcasts were included there, too. No static, though you couldn't hear the broadcasts on the ground unless you could see the VOR station.

If the weather was too bad to fly, as evidenced by a phone briefing or just a look out the window, we would listen to the scheduled broadcasts, carefully writing down each sequence report (now called METARs). The observations were made just before the hour unless there was a significant change in the weather, so it was the 15-after broadcast that was the most significant. The big question: Better than last hour or worse than last hour? The basics were important, too. Wind direction and velocity? Temperature and dew point? Watch the trends. If a front was involved, when did it pass a station to the west? How strong was the southerly flow ahead of the front and the northwesterly flow behind the front? Was the temperature change sudden and substantial? These things were clues to both the possibility of thunderstorms and the severity of any storms in the frontal zone. Most of us learned early on that, as today, you can't wish weather to be better than it wants to be. We learned a lot, too, as we tracked fronts and storms and contemplated their effect on surface weather.

Forward to Now

The reason I acquaint you with what may seem ancient aeronautical history is that there is today a hue and cry about the diminishing availability of those one-on-one briefings. The FAA instituted a weather dissemination system based on using personal computers and modems— DUAT, for direct user access terminal. That means we can get all the information that has been available to a flight service specialist and brief ourselves and file our flight plans. Actually this was all available before DUAT through various private services. The feeling is that the FAA will perhaps try to evolve into a system where it makes available the information and it becomes the lot of the pilot to do all the interpretation. In other words, there'll be no more one-on-one telephone briefings. If this happens, private weather services that offer, among other things, telephone briefings will no doubt spring up. But most pilots will probably learn to interpret the National Weather Service information that is made available through personal computers at home, at work, and at fixed-base operators. To be able to do this effectively, we will have to spend more time studying weather.

The effect of this on safety is open to question. A pilot who puts some effort into developing weather wisdom can probably do a better job of self-briefing than an FSS, only for the reason that a pilot should better understand his capabilities. It has always amused me that FSS specialists ask the type airplane before they will brief you. As mentioned earlier, this doesn't really matter—weather is weather and knows not if it is being penetrated by a 182 or a 747. What does matter is how weather-wise the pilot of the airplane happens to be. Is he well trained and current? What equipment does the airplane carry? If pilots will put the proper effort into learning the fine art of self-briefing, the system should actually be better. Certainly with a complete briefing from a computer we take off with a lot more information than is available in a five-minute telephone conversation.

Works Three Ways

Forecasts are educated guesses on what the weather might be. The National Weather Service (NWS) has several forecast products that we use. The area forecasts cover a wide area—six cover the contiguous United States—and are issued three times a day. The aerodrome forecasts (TAFs) cover individual airports and are issued four times a day. The convective outlooks are issued or amended hourly, three cover the U.S., and they project what existing or developing activity will do. These are issued even if they are negative statements. A convective outlook is issued twice a day for the possibility of thunderstorm development over the next 24 hours. Areas where severe activity is forecast are often shown in red on the TV weather shows. One thing to remember about all the forecast products is that they can work three ways. The weather might be better than forecast, it might be as forecast, or it might be worse than forecast. In using the products of the NWS that cover actual happenings and combining them with what we actually experience, we make our grade of the forecasts.

Real Time

The primary items of weather information, as opposed to forecasts, are found in METARs, radar reports, pilot reports, and convective SIGMETs. Regular SIGMETs may or may not be information—they cover the probability of severe icing, severe and extreme turbulence, and widespread dust storms, sandstorms, or volcanic ash lowering visibilities to less than three miles. In the case of icing and turbulence, they would be more forecast than anything else because in rapidly changing conditions even a pilot report of either becomes suspect after, say, 30 minutes.

In Order

The FAA and National Weather Service put much thought into developing a format for a briefing. There is logic to the way it is done and the way the various products are presented to us.

First we want word on conditions that might cause serious problems—including thunderstorms, ice, low ceilings, mountain obscuration, and turbulence. All these delights are warned of at the beginning of each of the six area forecasts, in the precautionary statements section. Here you will be referred to the AIRMETs that are in effect and what they cover. The details are in the AIRMETs themselves, filed at the end of the area forecast along with SIGMETs and thunderstorm discussions. If you ever wondered where some FSS folks come up with the gloom and doom that pervades some briefings, this is the place. Just remember that the forecaster is covering all the bases — when he includes a state in a precautionary section, he wants to make sure we take the precaution of going into all the available information to ensure none of the bad things are harassing our proposed route of flight. It is a flag.

After looking at the flight precautions, we read the synopsis describing the general weather situation to find the reason for any bad things. We can visualize this or draw it on a map. Hopefully, some idea was in mind before the briefing. A weather-wise pilot keeps up a continuing interest in how weather systems have been moving by watching the weather shows on TV. Just being aware of the wind direction and velocity on the surface at your location gives a clue to the weather. In the Northern Hemisphere put your back to the wind and point left toward the area of low pressure. Looking at the sky tells a lot, too. Low and fast-moving clouds from the south suggests a storm is coming because there's plenty of moisture and a low to the west.

Pig Latin
When we were kids we used to talk in what we fondly called pig Latin, or *ig-pay atin-lay* — we put the first letter of each word at the end and added an "ay." The National Weather Service seems to use a language of its own in writing forecasts. It's not pig Latin but an innovative set of contractions; a full list of them is available in *Aviation Weather Services* (Advisory Circular 00-45) and there are software programs that will decode some or all, but don't bet on every forecaster following the contractions to the letter. Even without the list you can do a pretty good job of reading all but the most technical discussions, which are usually found in the convective outlooks. Trying to pronounce the contractions may get you close to some of the whole words. Some are too brief and too technical for this — UVV for "upward vertical velocity" always comes to mind.

Briefings
There was an interesting weather system active as this was written, and in three days it gave the opportunity to use the computer system to look

at possible flights that would encounter fronts, icing, and thunderstorms and to evaluate the completeness of the information. The system was noteworthy because around St. Valentine's Day it brought record warmth to the east and record cold to the west with a substantial mess in between. All the weather forecasters remarked that it was a tough system to predict.

The first day, for a trip from New Orleans to Wichita, the precautionary section mentioned only turbulence. The synopsis from the Dallas/Fort Worth area forecast was: "CDFNT FROM MI TO SERN KS WWD TO OK PHNDL THEN NWWD. CDFNT MOVG SLOWLY SEWD. BY 05Z CDFNT FROM NY TO NERN AR ACRS SERN OK WWD INTO E/ CNTRL NM." So there was a cold front from Michigan to southeastern Kansas westward to the Oklahoma panhandle then northwestward, and the cold front was moving slowly southeastward. By 0500Z the cold front was expected to stretch from New York to northeastern Arkansas across southeastern Oklahoma and westward into east-central New Mexico. That is neither hard to read nor to visualize.

Where's the Low?

My first curiosity would relate to the low-pressure system that created the front. It wasn't mentioned on the DFW area forecast, and the Chicago area forecast, which would also be a part of a briefing, didn't have it either. So the low must be pretty far to the northeast, out of the Chicago area. That's why the front is moving slowly. If we want further confirmation of that, look at a METAR from a station on the northwest side of the front. Wichita was good VFR with wind from 360 at 15, or, shifting around almost parallel to the front. Cold air there, though—the temperature at Wichita dropped 18° in two hours.

It is interesting that the character of this weather system apparently changed overnight. On the very-early-morning TV reports they showed an area of forecast thunderstorms over Louisiana and Arkansas, which is why I called up a complete briefing. But that changed. No SIGMETs or convective SIGMETs; the only flight precaution was for moderate turbulence below 6,000 feet—a given with the surface wind gusting up to 30 knots in the stronger winds ahead of the front. No IFR conditions were suggested in the precautionary statements section and all the TAFs were for VFR conditions; the METARs agreed. So this trip would likely be fine, IFR or VFR, with maybe a few bumps, especially near and in the frontal zone. One interesting item: The Wichita TAF called for 4,000 overcast and a northeast wind at 15 knots toward the end of the period.

So the forecaster was anticipating that the cold front would stop; witness the northeast wind. The next step? Wait for a new low to form on the stationary front. Then you would get a new set of fronts. As we discussed earlier, exactly where and when the low forms is one of the more difficult predictions. The winds aloft forecast called for a continuing warm southwesterly flow above 6,000 feet, even after the cold frontal passage, so there would be a good moisture supply. The only sure bet was that things in and north of the frontal zone would probably get worse before they got better.

Icy Day

Go now to the next day, when the system had become more complex. Decisions on wintry days can be difficult because of the possibility of airframe icing. When flying airplanes without approved icing gear, we are, in effect, flying barefoot. Radar reports can tell us that thunderstorms are on the prowl, but there is nothing that tells us whether ice exists. Pilot reports can help, but they cover only one altitude, one location, one time in the past, and one pilot's opinion. So it takes a cautious approach, one where we plan a flight with multiple Plan B's. Or perhaps it is a wait for improvement. With this winter storm getting more active, the questions became more interesting and related to ice. Low-pressure waves appeared to be moving along the stationary front, a low appeared to be forming out west, and there was a big upper-level trough with the tip down into the desert southwest. In other words, things were about to get moving. The morning TV showed an icy band in the Ohio Valley and told of a possible major ice storm in Indiana and Illinois. The TV map reflected only expectations, though, because the storm system was still sorting itself out and getting organized.

Go Flying?

How would that have been to fly through?

With a briefing in hand for a trip to Kansas City from Morgantown, West Virginia, this proposed flight was one where we would take advantage of more features of the area forecast and use them in a different way than for the relatively simple flight. A lot of information had to be studied before a decision could be made. By poring through all those pages of information, I could fly knowing for sure that no stone was left unturned.

Starting with the area forecast, the first thing was an impressive list of precautions. On the Boston area forecast virtually every state was touched by one or more of the list of IFR, icing, turbulence, and mountain obscuration. This proposed flight was IFR, so the precaution on IFR conditions would be to make certain of minimums for the destination and alternate. Every state we would fly across was on the list. The one on icing also covered every state we'd fly across. (Some were on the Chicago area forecast—two or more area forecasts might have to be considered for a trip of any length.) The precaution for turbulence was only for West Virginia; mountain obscuration would have to be considered for West Virginia and Maryland.

Synopsis
For the next clue, to the synopsis: "CDFNT AT 06Z...INVOF CAR-ALB-PIT-CVG-FAM LN WITH NRN END ACLTG EWD...BCMG NRLY STNRY PIT WWD. FNTL SYS MOVG TO YSJ-AVP-PKB-LOU PSN BY 18Z...CONTG TO ACK-ACY-HNN-ARG PSN BY 00Z."

A couple of things. The forecaster is telling us that the north end of the front is accelerating eastward (ACLTG EWD).

Next, the location identifiers are going to be unknowns to many. Check the appendix in AC 00-45 for a list of these identifiers. But some of them don't really matter because they are far from the route of flight. Most of us wouldn't know that CAR is Caribou, Maine. More would know Albany, but for this flight what matters is the Pittsburgh–Cincinnati–Farmington, Missouri part (many of these identifiers have been changed since this flight). In going from Morgantown to Kansas City, we would be starting out south of the stationary front and going through it at an oblique angle. What does this mean? It means that if there is nastiness in the front, we'll be flying in that nastiness for quite a while.

IFR
With that in mind, let's consider the IFR precautions. Missouri was on the list, so the destination and an alternate had to be considered. The terminal forecast is the place to go for that. Let's say we'd be pulling into Kansas City after 17Z. The forecast: "3-FZRAPL OVC25." With that 2,500 overcast and occasional visibility restriction down to three miles, no alternate would be legally required. But we would sure want to know where better weather would likely be located if that forecast went sour. Because the forecaster was covering the possibility of waves developing on that front, there was no reasonably close place with a forecast substantially better than Kansas City. If the flight were to be undertaken, then, the enroute weather checking would have to be meticulous. At the first sign

that Kansas City would be worse than forecast, the nearest airport with minimums might become the best haven. The ground is a more pleasant place than the air to contemplate busted forecasts. If the decision were made to go, it would best be with full consideration of that occasional light freezing rain in the forecast.

The final IFR item to consider, along with mountain obscuration, relates to the available en route options in case a problem should arise. What would we be able to do if some problem arose with the airplane or equipment along the way? The TAF/METAR reports give good clues here. That day the lowest ceiling along the way was reported as 500 broken, two miles visibility in light rain and fog. Morgantown had 4,500 overcast, Parkersburg 2,700 overcast, both with excellent visibility, so the mountains were not obscured at the time of those observations.

The terminal forecasts along the way didn't call for anything worse than a chance of 500 overcast, so almost all the IFR airports along the way were forecast to be at or above minimums, leaving a lot of options. Looking westward, the first mention of freezing rain was in the forecast for Decatur, Illinois. If the flight were undertaken, the forecaster was betting that you wouldn't likely have freezing rain at the surface until in the vicinity of that airport.

Icicles

The final and most significant flight precaution to consider would be icing. There was no SIGMET, so no severe icing was expected. The freezing level portion of the area forecast put the level at 5,500 to 7,000 at Indianapolis; moving westward toward colder air, the freezing level was forecast at the surface by St. Louis and at 4,000 to 8,000 over the rest of Missouri and Indiana. Okay to start out, questionable later on. What about an inversion? The only two places to look for those are in pilot reports and temperature aloft forecasts. The wind forecasts over western Missouri called for a plus reading at 9,000 feet throughout the period — but only a plus one. The forecast temperature was at or below freezing above and below that level. The only pertinent pilot report, albeit an old one, told of light rime ice over Kansas City at 4,000 feet.

So what would be the verdict, sports fans? From all available information it certainly looked okay to start out IFR even without deice capability. But the trip would have to be flown with a willingness to retreat and land short of Kansas City unless actual conditions turned out to be somewhat better than forecast. Starting toward Kansas City in an airplane without deice capability and with a pilot determined to get there could result in a bad scene. We have to always be suspicious, but whenever the temperature is within a few degrees of freezing be especially wary. If it starts to get colder, now rather than later becomes a time for action.

Add Ol' Thor

When thunderstorms are added to the equation, the importance of the available information takes yet another blip upward. In the same weather system that we have been discussing, the action heated up the following day. An upper-level trough lent support to the surface action and the front started moving, if slowly. For the sake of exploration, let's look at a trip from Evansville, Indiana to Little Rock, Arkansas.

Looking first at the area forecasts (DFW and Chicago cover the route), we see flight precautions for IFR, icing, turbulence, and thunderstorms for the states over which we'd fly. I don't know about you, but thunderstorms always get my first interest, so the next jump would be to the convective SIGMETs. There were three valid ones. One covered an area in Louisiana, another an area in Louisiana and Texas, and then there was one for the route under consideration: "FROM 40 NNE DYR-20SSE MEM-50 NNW MLU-60NFSM-40NNE DYR AREA EMBDD TSTMS MOVG FROM 2320. TOPS TO 400." For our route, that line from 40 north of Fort Smith to 40 north-northeast of Dyersburg defines the problem. We would enter the area of embedded storms about halfway into the flight. Go around it? Using the Memphis and Monroe, Louisiana references, we see that Little Rock is pretty well surrounded by the area. To get there, the area would have to be entered.

Radar

The SIGMET just defines the area. To get the amount of coverage in the area, we have to go to the radar reports. The one from Little Rock (identifier LZK for weather radar reporting purposes) brings rather bad news: "LZK 1325 AREA 8TRWX/NC 47/220 86/160 207/105 295/135 331/160 C2240 MT 380 AT 307/53 TROP 393." (The Little Rock reports also showed a new area to the south with tops to 420.) So at 1325Z the area was 80 percent covered by intense thunderstorms (the X equals intense; only an XX, extreme, is worse), no change in intensity, and 47 degrees at 220 nautical miles tells the tale in the direction affecting the proposed route of flight. Eighty percent coverage is just about solid, which is what this area looked like on the morning TV shows—solid. Even with a radar and a Stormscope, this flight would be highly questionable.

Still, it is worth looking at the rest of the information. After the convective SIGMETs is the four-hour outlook. This one described the conditions that were causing the activity and noted that satellite imagery showed the cloud tops had warmed during the past hour, indicating a weakening trend. Why? Warmer tops are an indication of less instability—remember, it is cold air over warm that causes instability. However, they expected redevelopment because of increased upper-level support moving into the area and included the possibility of strong and possibly severe thunderstorms toward the end of the four-hour period. These convective outlooks are as technical as the information gets, but they are not that hard to read; even if you don't get all the contractions, the message is fairly clear.

The actual weather along the way ranged from 1,500 broken and 25,000 overcast to 400- to 500-foot ceilings at some of the stations. A thunderstorm was reported to have passed Jonesboro, Arkansas, 25 minutes before the observation, and one was in progress at Paducah, Kentucky.

The only good news was that the freezing level was up around 10,000 feet, so ice would not be a problem for a low-altitude IFR flight.

The report of the activity weakening because of warming cloud tops might send a less-than-timid soul out to have a look. But it would have to be a careful look, flown with a complete willingness to land along the way should the thunderstorms be too close together for safe passage— almost a certainty with 80 percent coverage.

Wind and Turbulence

Those reports all involved a slow-moving weather system. A late-November self-briefing for an actual trip from Maryland to Springfield, Ohio, was for a fast-mover and required as much interpretation as the complex slow systems. Actually, the interpretation started the night before as strong winds in advance of and then behind a cold front rattled the shutters and did such mean things as rip hangar doors off at the airport and spread debris all over the place. Before I called up a briefing on my computer, I knew of the primary factor to be resolved—wind. All the early-morning weather people on TV were busy explaining why the wind was so strong, and there were reports of damage at airports and elsewhere. There were whitecaps on the lake in front of my house, and scattered to broken clouds were moving briskly from northwest to southeast.

SIGMETs (as opposed to convective SIGMETs which cover thunderstorms) are issued whenever severe or extreme turbulence, severe icing, or widespread duststorms or sandstorms (or volcanic ash) reducing visibility to below three miles are expected. That day it would be turbulence, if anything; the severe version of that is said to toss things about, force you against the straps, and result in large changes in attitude and/or altitude, usually causing large changes in indicated airspeed. They add: "Aircraft may be momentarily out of control." Not my idea of a pleasant way to spend the morning, so the first thing I went for in the briefing was the SIGMET section. There it was: "FQT MDT TO OCNL SVR TURBC BLO 100 DUE TO STG LOW LVL WNDS WITH LCL STG UDDF OVR AND NR MTNS. LLWS PSBL. CONDS CONTG BYD 1415Z." (UDDF means up- and downdrafts; LLWS is low-level wind shear.)

With such a forecast, it becomes the pilot's job to not fly, to bear the turbulence, or to convince himself that the turbulence will not be quite so bad. That morning I had wanted to fly, but I didn't want to get pounded by severe turbulence. So I went looking for something that might suggest the air wouldn't be quite so lumpy.

I found what I was looking for in the reports of actual conditions, the METARs (at that time they were called "sequence reports"). They looked good. Every reporting station within 50 miles showed diminishing surface winds. Baltimore, fairly close by, had dropped from 43 to 24 knots. The strongest wind gust anywhere around or along the route was 30 knots, not a number that reflects severe turbulence. All the velocities were below forecast values. I made the decision to go, encountered some light to moderate turbulence on the climb, and then enjoyed smooth air at 10,000 feet. Further evidence that the strong wind was abating more rapidly than expected came from the flight time. It was 38 minutes under that predicted by the computerized flight plan. A colleague who had planned a flight that morning got his briefing from a Flight Service Station and opted not to fly. He was disappointed but should not have been. He made a good conservative decision based on what he was told.

Big Question

For most pilots, the biggest question that will arise from self-briefing relates to what action you take if you are not sure. It is really no different than when an FSS person does the briefing and you can always call for a second opinion if you so desire. The information is the same, only the interpretation is different. And, hey, nobody ever said making decisions wasn't a lonesome business. The key to some of those flights was that it might have been okay to start the flight, but it would be continued on to the destination only if certain things were possible. In a continuing study of general aviation accidents, I have found that although a few airplanes do come to grief in the first phases of flight, those are more often related to pilot technique or mechanical problems than to unflyable weather. More often, when a pilot trying to fly VFR loses the battle with weather, it is related to continuing VFR into adverse weather; IFR it is sometimes enroute weather (ice or thunderstorms) but more often it is low weather for an approach and a duck-under that results in an arrival before the runway is reached. That is pilot technique, too, but it is weather-related pilot technique. Back to the question: If a pilot is not sure, or leaves thinking he can make it through the conditions that exist but with some doubt, then the best deal is to decide not to start out just now. The only absolutely sure thing about weather is that it will change.

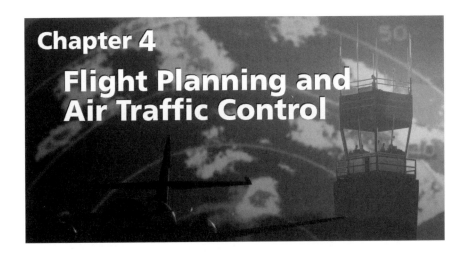

Chapter 4
Flight Planning and Air Traffic Control

To many pilots, the air traffic control system is a mysterious "they." It is a voice on the other end of the radio, a voice that a lot of us find friendly but some pilots feel is adversarial. The reason the latter feel that way is because what the pilot wants to do, and what air traffic control can allow that pilot to do, are not always one in the same. So the need is for an understanding of the system. What is possible and what is not?

There are three ways to deal with the traffic control system. When flying IFR, we have only to plan a route; file a flight plan, and do what they say. Participation in the IFR system takes care of dealing with regulated airspace. If flying pure VFR—that is, not talking to anyone on the radio—we have to plan to avoid regulated airspace (Class B, Class C, restricted areas, and the like) and not stray into space where the minimum equipment requirement is unmet. The third way is flying VFR in contact with air traffic control. This is informal, and any service given by a controller, except where we are required to talk to him because it is regulated airspace, may or may not be useful or available.

IFR

Pilots have wondered why we file routes for IFR when the system apparently sends us along the way more in accordance with its plan than our plan. But if you look at it in retrospect, most of the clearances you get are "as filed" with perhaps some exceptions, usually on departure and arrival. With the FAA allowing us to file our flight plans through a computer, we learn that their system isn't all that slick, too. On one flight plan I apparently made an error in a Victor airway. The computer told me so. When there is an error in routing the best thing to do is file direct between the two points along the way, which I did. Except instead of typing in ANX, for Napoleon VORTAC near Kansas City, I missed by one key and entered AND. The FAA computer was perfectly happy with a flight plan that went from Macon, Missouri to Anderson, South Carolina and then back to Olathe, Kansas. Some dogleg, and it was not corrected until just before we reached Macon, when the center controller called and said, "Something is not right here."

Who Is Out There?

It is often said that the air traffic control system was designed for the airlines; general aviation and military take what is left. There is some truth to this. The airlines constitute the majority of the usage of the IFR system. Just considering towers, where both VFR and IFR count, general aviation is the predominant user. In general aviation, single-engine airplanes, piston twins, and turbine-powered airplanes use IFR about equally.

Perhaps, though, it is simplistic to say the system was designed for airlines. Rather, it was designed to connect major airports. In the Dallas/ Fort Worth area, the big DFW airport dictated the airspace design. To accommodate Love Field, they carved out airspace beneath what used to feed DFW and handle its departures. The Class B airspace areas are all based on large busy airports, and Class C airspace areas serve not quite so big and busy airports. The purpose of the system is not only to provide separation between aircraft flying IFR and to manage the flow of traffic in busy areas, it is also to prevent collisions between VFR and IFR aircraft by procedural separation, or, said another way, airspace restrictions such as Class B areas.

When we are flying IFR, a route or altitude that seems arbitrary is usually assigned for procedural reasons. By managing the way traffic flows, the workload on individual controllers can be reduced. This became more noticeable after the air traffic controllers' strike in the early 80s. Faced with training and putting on duty a virtual new set of controllers, the FAA had to rely more on procedures that were followed without exception. The idea that you had to always fly the prescribed pretzel regardless of traffic grated on the nerves of a lot of pilots. But it worked, and still does.

Example

The procedural business has had some benefits, too, in the usual bad news/good news scenario. When I was based at Trenton, New Jersey, for years we came in from the west over the Bucks (formerly Bucktown) intersection, the same route followed by airplanes flying into Philadelphia International Airport. Separation was usually by altitude—we'd come in 1,000 feet below the aircraft headed for Philly. If they got fouled up and holding commenced for Philly, it affected traffic going to Trenton. I was never particularly comfortable with this. For one thing, watching a string of jets whiz directly overhead, 1,000 feet above, made you mull what might happen if there was the slightest error and one of them descended. It was the kind of deal where you listened carefully, making sure everyone was doing the right thing. The other bad scene was the holding pattern. Under certain atmospheric conditions wake turbulence can settle a full 1,000 feet while still maintaining a bit of a punch—nothing you can't handle but, still, poking around inside a cloud with a little twisting motion in there to make things interesting was never my idea of a good way to spend the afternoon. I had a couple of wake turbulence jolts in the vicinity of Bucks and in general found it an uncomfortable location.

All this changed. They moved the inbound route to the satellite airports to the north, over Pottstown, Pennsylvania. Gone were the holding patterns and the jets overhead. No more wake turbulence. The bad news is that we were down around 5,000 feet 75 or 80 miles from Trenton. That was uncomfortable in low-level turbulence. The turbine airplane operators sure didn't like it because of the high fuel flows down low. It did, though, provide a measure of procedural separation and had to be viewed as an improvement.

Second-Class Citizen?

Does all this mean that a pilot flying from Clermont Airport near Cincinnati, Ohio, to Carroll County Airport near Baltimore, Maryland, is a second-class citizen in the eyes of air traffic control? No way. Nor, though, is there a way to tailor airspace to give the pilot flying between these two points an optimum flight profile. Leaving Clermont is pretty straightforward, though it does require a bit of a dogleg because of a restricted area to the east. You can file either north or south of the restricted area and expect to get pretty much what you file for the en route portion of the flight.

Getting into Carroll, which is northeast of Dulles, north of Washington National, and northwest of Baltimore, requires peacemaking with the steady streams of traffic at those three busy airports. Even though they are 30 to 40 miles away from Carroll, you are using airspace that is under the influence of the procedures used to move traffic to and from the busy terminals. The price paid is a descent to 7,000 or 9,000 by the time Martinsburg, 44 miles out, is reached. Then, if it is to be a visual approach, the airport is usually right below before you are cleared out of 5,000 feet. That procedure makes it work for light airplanes headed for this airport. Often we can sit in judgment of the procedures used and decide they don't make sense; usually when you look into the matter and come to understand why they do it like they do, it isn't easy to offer an alternative.

Getting by New York is another example of procedural separation that appears to some as discriminatory. There are two choices when headed from points southwest to points northeast of the New York metropolitan area. Either pass by 40 miles to the southeast, pretty far out to sea, or about 60 miles to the northwest. This can add a lot of miles to some trips. What this does for the controllers is make low-altitude enroute IFR traffic no factor as they work the continuing puzzle of how to move high volumes of traffic at LaGuardia, Kennedy, Newark, Teterboro, and White Plains, which are in close proximity to one another. The good news here is that we can fly IFR into those major airports if necessary (and if you don't mind the high landing fee at some); the bad news is that if we are flying past the area, we have to go pretty far out of the way, but even that has a benefit. If the traffic in the New York area is fouled up, we can usually slip by unmolested.

Some pilots defeat this in VFR conditions by flying over the top of the New York Class B airspace VFR without talking to anyone. It is a perfectly legal procedure (assuming there aren't any extenuating circumstances conveyed in a NOTAM). I tried it once, staying very alert and watching for other traffic. It would sure make the trip from New Jersey to Boston quicker. There is not much traffic right above the airports in question, but I was amazed at how much traffic there was in the area as a whole. A lot of fast jets—it almost seemed as if someone had launched many long shiny missiles at my little green airplane. It was one of those deals where I promised, if I escaped successfully, never ever to do it again.

Good Treatment

The easiest way to go to a major airport is IFR, and in the course of flying to Kennedy I have never had a delay or anything other than first-class handling. The ultimate in convenience for a light airplane came one day when I took Captain John Cook of British Airways and his wife Joy, to Cook's left front seat in Concorde. On request, they let us land on 31 Right, a turnoff that puts you close to the British Airways terminal, where they allow general aviation parking on special request and with permission under certain restrictions. Anyway, we parked next to Concorde and I treasure the photo of John and Joy standing by my 210 with Concorde looming majestically in the background. It's also fun that they posted N40RC as an arrival on the TV monitor in the dispatch office.

Flexibility

The bottom line on all this is that in return for the flexibility of having an airplane that does equally well at Carroll County and Kennedy, we have to conform to the system that makes it all possible. It helps if we have an understanding of why some things are true, and how we can "work" the system to best advantage.

To begin, an IFR flight plan causes the electronic transfer of information to all the facilities that will be handling the flight. This doesn't include everything on the flight plan, just what is necessary for traffic control purposes. The controller needs to know that it is a Cessna 210 with approved area navigation gear; he doesn't need to know that the airplane is green and white.

When we call clearance delivery, or whomever we get the clearance from, the clearance should be there ready to go at any time from about 30 minutes prior to the proposed time off to two hours after the proposed time off. After two hours the flight plan "times out," as they say.

Just getting the clearance doesn't mean you can fly away. The airplane has to be released into the system. If you are at an airport where the tower and radar-equipped approach control are located, you probably won't be aware of this process. If you are at a satellite airport with a tower, the tower will probably say something like "Hold for IFR release." Then he'll ring up the approach control with responsibility for the airport, tell them you are ready, and they will give a release or a time to expect one. The same procedure would be true with an FSS; the release might come from an approach control or an air route traffic control center. If there is a remote clearance delivery frequency at an uncontrolled airport, you'd get the clearance and the release on that frequency. Another way to do it at uncontrolled airports is on the phone. You get the clearance, and they give a release that is valid for a certain time period. This seems awkward when you think about it, but it works well at most locations. That is the way it used to be done at Carroll, and after two years of flying there the only substantial delays I had resulted from inbound aircraft shooting multiple approaches when the weather at the airport precluded landing. Getting a clearance on the phone wasn't comfortable in the winter, or when it was raining if the office wasn't open yet. Then we had to use an outdoor pay phone. Getting a clearance with a void time requires some planning because they prefer the time to be as soon as you can manage and they usually give a five-minute window, such as valid at 45 past the hour and void at 50. I got the airplane ready, told the passengers to be ready for an immediate departure (or load them in advance), preheated the engine in the winter, had all the charts out and in place, and ran the checklist right up to the part about starting the engine. That way I could get off quickly without rushing anything.

I have long since moved my airplane to Hagerstown, Maryland. This airport has a control tower and that makes IFR departures a lot simpler.

As They Come

Most departure releases come from a terminal radar control facility — approach and departure control — simply because most airports fall under the influence of one of these facilities. If the airport is out in the open spaces the release would come from the air route traffic control center. The center, which handles en route and some departure and approach traffic at airports where there is no approach control, basically takes airplanes as they come. If a center is having a problem with too many airplanes or with weather, it might put an "in-trail" restriction on departures. Instead of the minimum five-mile separation, the Center might insist on, for example, 20-mile in-trail spacing. That is not something pilots really need to understand, but if you are sitting there waiting to go, you might as well know why things are bottled up.

There are all manner of letters of agreement between abutting approach controls, and between other facilities that spell out how they will move airplanes between airspace controlled by the two facilities. And although you can occasionally get an exception to a route that is spelled out in a letter of agreement, you'll usually fly the agreed-on route and altitude. Especially in major areas, they have altitudes as well as routes that are used for certain purposes. If you need a different altitude, maybe because of ice or turbulence, the controller will likely try to get it for you, but, because of other traffic, it simply might not be possible. A lower altitude, for example, might be in use for traffic going to another area, at right angles to your flight path. By taking some of the flexibility out of busy areas, they have been able to increase the amount of traffic handled without a corresponding increase in controller staffing.

Another term we hear is "gate." This isn't the gate at which you park but rather a departure or arrival gate in the airspace. This is just a form of procedural separation. At many busy areas they use a four-corner system for arrivals. Everyone comes in through one of the four corner gates — faster traffic higher than slower traffic. Departures go out between the four corners. If you have never been to a terminal radar control facility, you should go for a visit. After you watch them work the traffic in and out, you'll probably never again bellyache about an arrival or departure that appears to be a bit out of the way.

Terrain and Obstruction Clearance

On a published instrument approach, following the procedures keeps the airplane safely above obstructions and terrain until such time as the pilot sees the runway and makes a visual descent. Departures are not quite as simple.

Obstacle clearance for an IFR departure is based on the airplane being 35 feet high over the departure end of the runway and on it climbing 200 feet for each mile flown. It is assumed the airplane will climb 400 feet before making any turn. The safety margin is in the FAA using a 152-foot-per-nautical-mile slope in considering obstacles. Starting at that 35 feet above the departure end of the runway, if no obstacles poke into that 152-foot-per-nautical-mile slope, then no IFR departure procedure is published for the airport or runway. If one is published, it will be flagged on the NACO approach chart and detailed elsewhere in the book. Jeppesen presents them right on the approach chart. If a published departure procedure requires a climb of more than 200 feet for each nautical mile flown, the amount of climbing ability required will be stated on the procedure.

The clearance provided by this standard isn't great. It starts 35 feet high over the end of the runway and gives an additional 48 feet of clearance for each mile you fly. Two hundred feet per nautical mile isn't a lot of climb—at 120 knots it would be but 400 feet per minute, but climbing with a tailwind component on a warm day could do the dirty trick.

There is a kicker, too. If you are leaving from an airport located in Class G airspace, only those obstacles that extend up into controlled airspace are considered. Until you enter controlled airspace, usually at an altitude of 700 feet above the ground, it is up to you to arrange for obstacle clearance. The airfield at Carroll County, Maryland, is a perfect example of why we should do a survey before departing, to be aware of obstructions and any requirement for a rate of climb better than that 200 feet per nautical mile.

The most pertinent obstruction at Carroll is an antenna on a small ridge line two miles from the airport, close to the extended centerline of Runway 16. It is 1,137 feet MSL, 348 feet above the airport elevation. The 200-foot-per-mile climb minimum beginning at 35 feet over the departure end of Runway 16 would clear it by only 87 feet. That's not much. If it reached up into controlled airspace, there would have to be a published departure that addressed this particular obstacle. As it is, the chart says that after departing on Runway 16 you have to climb at a rate of 210 feet per nautical mile to 1,300 feet.

At Hagerstown we have a published obstacle departure procedure off Runway 9 that specifies a climbing left turn to a heading of 040 to 2,500 feet, to intercept the Hagerstown 084 radial before departing as cleared. That really has nothing to do with obstacles, it is to make sure you turn away from the regulated airspace at Camp David, to the east. In actual practice almost every clearance out of the airport is to the Hagerstown VOR, which is to the west and they expect you to turn toward the VOR instead of flying the published procedure when you depart from Runway 9.

En Route

It is when we are en route IFR that we find the greatest flexibility in the system. If you fly the same routes on a regular basis, it is fun to learn to work the kinks out of trips. One that I finally figured out after years was from the southwest into New Jersey. Flying my P210 in the low Flight Levels, I always had a problem with Washington Center in the vicinity of Elkins, West Virginia. They would usually offer some horrendous vector out of the way, to keep me clear of traffic that was descending to land at Dulles. So I got the chart out and figured a route that was only slightly out of the way but that avoided Washington Center's airspace entirely — direct from Bowling Green, Kentucky, to Henderson, West Virginia, and then to Harrisburg, Pennsylvania which was the feeder fix on into Trenton. That worked like a charm. No vectors and no conflict with other traffic. That, though, is the type of thing you have to work out yourself. The system doesn't seem to anticipate conflicts in advance, and it isn't until you get to the controller who has to solve the problem that it becomes apparent you would be better off somewhere else. We have to be aware that controllers are really good only at solving problems in the airspace for which they are responsible. When one controller gets you to the edge of his sector and hands you off to another controller, the new one might be marching to a different drummer and might have to give you a big turn.

Direct

The business about flying direct to somewhere is important, especially with the proliferation of GPS navigators. It is really easier for the controllers if you do the direct business when you file the flight plan rather than ask for it once aloft. Otherwise, when you request direct en route they might have to coordinate with controllers on down the line. Generally, if you use one high-altitude VORTAC (the ones with Jet airways running over them) within each center's area of authority, it'll work well in their computer. If you want to try direct to an airport out of the center area in which you file, use both the airport identifier and the lat-long for the airport. Forget about filing direct to an airport in a busy area. You are going to have to fly the prescribed arrival, so the best thing is to file to a VORTAC they use as a feeder fix. If there is a standard terminal arrival route for the area, use one of the primary feeder VORTACs in the STAR.

Beehives

If we do our IFR business at lower altitudes we often have to deal with approach control facilities when en route because they control all the airspace up to, say, 10,000 feet and out 30 or more miles from the terminal. Some control more space, some less, but if we are going to motor through their air we are going to be under their control. This is no big problem, but if the area is a busy one, perhaps one of the hubs that has sprung up since deregulation, the routing might become convoluted. When they have a "push" at a hub, even an airport like Raleigh–Durham, North Carolina, can become congested. There is not much pilots can do about this, other than be patient, and the solution to a detour in a direction you don't want to go is not always the obvious one, as I found out at Raleigh–Durham one day.

There had been a series of tornadoes early in the morning of the day I wanted to pass by Raleigh–Durham at about noontime. The raunchy weather had all moved off to the east, but, as is so often the case, there was a lot of wind shear turbulence in the clouds behind the front. It was definitely very uncomfortable flying. My original plan was to stop at Florence, South Carolina, for fuel. When I came under the influence of Raleigh–Durham approach control they gave me a vector heading directly to the south for traffic. I wanted to go west, to get farther away from the front and to hopefully moderate the pounding I was taking from the wind shear turbulence. No way. No altitude change either. If I was going to Florence, all they had available was that one route and that one altitude because of all the airline traffic that was coming and going at the hub. I resigned myself to the turbulence, but then it got worse. The solution? I told them I would like to land at Raleigh–Durham. That eliminated the detour to the south and allowed me to file a new flight plan along a route that would be less turbulent. The object of going to Florence for fuel had been to avoid the busier airport, but the desirability of that plan faded fast. It probably made no more workload for the controllers for me to land at the busy airport as opposed to being vectored through the area.

Handoffs

As pilots, we like to change altitudes often. Controllers who don't fly airplanes probably have a hard time understanding why pilots like to go up and down like a bunch of yo-yos, but we do like to seek smoother air, better tailwinds, lighter headwinds, less ice, or a better view. It is simply not always possible to determine the best altitude when planning a flight. There are some simple rules for us to follow to make the altitude change routine easier.

If we ask for an altitude change right after being handed off from one controller to another, it is not likely to come through right away. Why? Because handoffs are usually made before you actually reach the airspace "owned" by the new controller. You are still in the old guy's air and the new guy can't give you a new altitude until you reach his air. So if you ask for lower, he is likely to reply with something like "Lower in five miles." The controller could coordinate with the previous one and get the lower altitude for you immediately, and would do so if there was some urgent reason and no traffic conflict. But things go more smoothly for everyone if you wait a few minutes before requesting a change from a new controller. It is all a matter of planning.

IFR Sums

IFR planning is really quite simple. And the reward of good planning is a flight that goes as expected. One example of this came on a flight that I flew occasionally, from my base in Carroll County, Maryland to Asheville, North Carolina, or Atlanta. I knew that there were two routes available from that base in Maryland to the southwest. One went to the west of the Washington area. I flew straight west to Martinsburg, West Virginia, then southwest over Montebello and Lynchburg in Virginia, and then on toward the destination. The other route went over Baltimore and southeast for a few miles before cutting back toward the southwest. In my advance planning I always noted the flight plan for both routes on the outbound and inbound forms because the weather had a lot to do with which way I wound up going. The western route was over the mountains, and if there was a strong westerly flow, the up- and downdrafts were a big factor and the turbulence often extended up well above 10,000 feet. If a front had passed through, though, the eastern route might put the airplane in more frontal weather for longer. So an important part of the plan was the alternate plan. One place I usually missed was on the route. Always, when I was headed from Martinsburg down the airway to Montebello, I decided that I wanted to go direct to Barretts Mountain or Sugarloaf VORTAC in North Carolina. Often when this was requested once en route, the controller was not able to approve it. But if I remembered to file it that way—direct from Martinsburg to Barretts Mountain or Sugarloaf—it was approved even though neither of those VORTACs was a high-altitude facility. That is just one example of how a good plan can make an IFR flight easier for the pilot and the air traffic controllers.

VFR Doings

If we want to fly pure VFR—that is, not talking to anyone and without a transponder in the aircraft—the planning burden is heavy but it surely isn't impossible. What is unlikely is being able to draw a straight line from here to there in any of the heavily populated areas. For example, if you wanted to fly from Frederick, Maryland to Franklin, Virginia, you would have to make a big detour to the west to stay out of the 30-nautical mile Mode C requirement rule for the Washington area. Once by, there are numerous landmarks and highways to use in missing some restricted areas. This would take you north of the Richmond Class C airspace. Then you'd have to mind the Class D airspace at Fort Eustis as you flew on into Franklin. The flight would have to be flown below 10,000 feet because Mode C is required above that level. At least that was how it would have been legal when this was written.

The key to staying out of VFR airspace trouble is to use only easily identifiable landmarks. We used to practice a form of this when flying VFR from Teterboro to Trenton, both in New Jersey and both very much under the influence of the New York Class B airspace. On busy days it was simply better to roll your own rather than courting the favor of an overworked air traffic control system. You could fly straight west to a ridge, then follow that ridge line to the southwest, going very close to a couple of reservoir ponds and a golf course and staying low to be beneath a fold of fat on the New York Class B airspace and navigating precisely to go through a quite narrow gap between the Class B airspace and the Morristown, New Jersey, Class D airspace. We used to joke that the route was available only to airplanes with a wingspan of 50 feet or less—otherwise you couldn't get through legally. Many airplanes used this because it was a fairly direct route, it had good landmarks, and it kept the airplane out of regulated airspace—then. Now you need a Mode C transponder, which hopefully most airplanes carry, to fly the route.

Scud Run

In a congested area, VFR navigation is done much the way it was when scud running was more socially acceptable and, like night cross-country, was taught before the advent of radio navigation in light airplanes. In both cases the technique was one of taking things one step at a time. A route that followed things—roads, railroads, or rivers—was carefully planned. Then landmarks along the route were identified and studied. The drill was, on reaching one landmark, to not proceed unless you could actually see the next landmark. If it couldn't be seen it was time to turn and retrace steps. In scud running, this could still be hazardous because of the temptation to go just a little farther for perhaps a view of that next checkpoint or because of worsening weather behind. A lot of airplanes were lost to the practice and it is hardly recommended. At night, in good weather, it worked much better. I have flown many a mile at night with no radio and no panel or landing lights—all the airplane had were navigation lights—flying down highways from one town to the next. It was often tempting to strike out cross-country along a more direct route, but I had been convinced early on that being unsure of position at night, with nothing but a map on board, is one of life's less desirable sensations. The key was to have a plan that ensured you'd not become misplaced. That same key works when it comes to visual navigation around regulated airspace.

Radio Flight Rules

I've always thought that the "other" form of VFR navigation through the airspace system should be called radio flight rules. The airplane is in visual conditions, but it is navigated with information from the avionics equipment, and the flight might be in contact with air traffic control all along the way. If there is a drawback to this it is in the pilot becoming complacent about other traffic. It is all too easy to become mesmerized by the array of information on the instrument panel—at the expense of that outside scan. And just being in contact with air traffic control and on "flight following" is no assurance of traffic separation. The controller will probably call all pertinent IFR traffic to you, but he might not even see VFR traffic.

The primary planning requirement here is to make certain you know, for example, how far you can go toward Class D, C, and B airspace, for example, before you make contact with the tower or approach control in the case of the Class D airspace, or, get a Class B clearance in the case of a Class B airspace area. It is also important to consider all factors. At Dallas/Fort Worth, for example, they used to have a habit of clearing you out of the Class B area to the northeast; once out you had precious little time to contact Addison tower before penetrating their Class D airspace. If all that came as a surprise, you'd be at a distinct disadvantage.

Corridors

I mentioned our little corridor in New Jersey earlier. There have been official corridors through Class B areas in New York, Los Angeles, and in the Washington Class B airspace. Atlanta's Class B airspace had one years ago but it was removed because of problems. Someone said that the vertical limits of the corridor corresponded to the altitudes where summertime cumulus form in the area, making the corridor difficult to use. For the ones that are left, the pilot requirement is to study the corridor procedures before using the corridor.

Planning is Everything

So in dealing with air traffic control, planning is everything. A bit of understanding helps too, and nothing helps the understanding more than some direct contact with the people who run the system. A visit to a tower, approach control facility, or center is an enlightening experience. And if once you make the plan they don't go along with it, take it with understanding. There's just no way they can let every pilot do it his way.

Chapter 5
Reading the Signs

Briefed on the weather, with a good flight plan, with the airplane carefully preflighted, and the checklist used to perfection, a pilot is ready to convert all this into transportation from here to there or, in the case of a local flight, from here to here. All the forecasts will be converted to "nowcasts," the meteorologist's jargon for forecasting in the very short term. What will happen next? How will it work? How do we keep from doing something risky? Certainly it is in the minutes before takeoff, when we are looking at the airplane and the sky, and assessing our ability in relation to the conditions, that the opportunity exists to avoid a risky departure. Whether it's something the weather is doing to the airplane, such as depositing ice on the flying surfaces, or something the weather is doing around the airport, such as creating lightning and thunder and wind, or something suspicious on the airplane, or some ill feeling on the part of the pilot, we have to read the signs and make the right decisions.

Example

A Pan American 727 crashed just after takeoff in New Orleans in the '80s. A thunderstorm moving toward the airport as the crew prepared to depart, and the resulting wind shifts caused the wind shear alarms to sound. It was quite obvious that the storm was there—it was a typical New Orleans summer day and the storm was not hidden by other clouds. Despite the occurrence of the storm, air commerce was being conducted at the airport, though different pilots had different ideas on how to deal with it and all didn't want to use the same runway. Against this background, the Pan American crew taxied out and prepared for departure.

The flight was to be a long one, and the airplane was quite heavy with a good passenger load. As they were taxiing, the crew had a discussion about the best procedures to use for takeoff and how they would handle an engine failure. While they were taxiing to Runway 10, other aircraft were requesting Runway 28, and at one point the tower read off the wind from three locations on the field: From 060 degrees at the tower, 330 at the northeast corner of the field, and 130 degrees at the northwest quadrant. Clearly it was a variable wind. Perhaps some pilots requested Runway 28 because it looked as if the initial climb path would be clear of the cell, whereas that didn't look as certain to the east. Or perhaps they were looking at a windsock.

The 727 took off on Runway 10 but remained airborne for only about 30 seconds, then the flight became one of the infamous "microburst" accidents. The strength of the downdraft combined with an increasing tailwind to simply best the performance capability of the heavily laden 727.

How do you avoid making a bad decision about weather before a takeoff? Perhaps the best way is to mentally dissect the atmosphere you are about to fly through. For example, if there is a thunderstorm to the northeast of the airport, approaching the airport, what factors must be considered? First, when a storm is moving toward a westerly direction, it might be stronger than the average summer thunderbumper — certainly it will be stronger on the west side. Westerly-moving storms are not moving with the standard west-to-east upper winds. Instead, they are within an air mass, an unstable air mass with light or calm winds aloft, and are just feeding on moisture. The downdraft from the storm would start to fan out near the surface, becoming a variable surface wind, and would have to be considered when making the decision that it is safe to take off. Quite pertinent to the southwesterly movement of this storm is that we are used to the outflow from a storm, the first gust, being strongest on the east side of a storm, the direction toward which most move. In this case it was likely on the west side.

The downdraft part of an encounter like this is bad, but at low altitude the shifting wind probably has the most effect on performance. When an airplane is lost in such conditions it is because it flies from an area with a headwind, to no wind, to a downwind condition. This results in a loss of performance because of an increasing tailwind, as discussed earlier.

A classic example of an increasing tailwind's effect on performance came in another 727 accident, in Denver some years ago. In this case the cell appeared to be large but weak and based at a rather high altitude. This 727 rotated normally and then flew into an increasing tailwind with the airspeed on 157 knots to begin. In five seconds the airspeed had decayed to 116 knots, too little to sustain flight. It was estimated that the 727 flew from an area with a 10- to 20-knot headwind into one where the airplane had a 60- to 90-knot tailwind component. In this case the storm that created the downdraft and shifting wind didn't look strong and, in fact, it dissipated shortly after the accident. Had there been a low-level wind shear system at the airport at the time, there would have been some warning. (The LLWS system is nothing more than wind-sensing devices on the surface at various locations around the airport. The wind shear warning comes when there is a substantial difference in wind at two of the locations.)

In both these cases it is clear that the report of wind from the tower location only, or a look at a windsock, doesn't tell us what we'll be flying through in those first minutes after takeoff. That is a judgment we have to make, based on an intelligent assessment of conditions. The flow out of a thunderstorm is not hard to visualize, and to avoid problems we have to make the extra effort to fully visualize the flow.

Traffic Control

If there are thunderstorms in the area and we do opt to fly, it has to be with a full understanding that there is not as much latitude for the air traffic controller to use in allowing deviations in a terminal area. If he has traffic, he has to keep it relatively close to the normal flow patterns. That is why, in widespread thunderstorm outbreaks, they have closed some of the major terminal areas to arrivals and departures. In New York, for example, the airspace divisions are so complex that it simply would not be possible to maintain any semblance of a normal traffic flow were every arriving and departing pilot is requesting a deviation around weather.

This came on an arrival but I remember one stormy evening virtually having the New York area to myself. When I requested a deviation, the controller said something to the effect that I could do whatever I wanted—nobody else was out there. I took the hint and went to the nearest airport and landed.

Whether arriving or departing, dealing with thunderstorms close to the airport is nothing like dealing with them en route, where we have wide latitude to deviate and, on a long trip, can go a hundred or more miles out of the way in order to stay out of the storms. In the terminal area, we are flying into or out of a funnel.

Thunderstorms are dynamic—that's why we had best stay out of them—and any check of conditions an hour before flight leaves us with a picture that might be out-of-date for takeoff. So it is a good idea to make a last-minute check with whatever means are at hand. If you have a Stormscope, does it indicate activity in the departure area? Airborne radar can be used on the ground (safely away from people and objects) to scan for activity in the departure path. You can ask ground control to ask departure control what it looks like out there. If you are at an uncontrolled field but can talk to the center or to a flight service station, you might query them for information on how it looks now—because now is what counts. Is there a convective SIGMET out for the area? These do an excellent job of defining thunderstorm problems. The nice thing about doing this before you take off is that a very simple and safe solution is always available. Wait. Once the airplane is in the air, you have to get it back on the ground before that safe and simple solution is again available.

An uplink of the National Weather Service radar is now available if the airplane has a multifunction display and the appropriate receiver. This can be very valuable but isn't something to be used for the penetration of an area of weather. Rather, it is like the first Stormscopes and is valuable mainly for the avoidance of areas of weather.

Bad Weather Takeoffs
The first weather consideration on departure is certainly not always a thunderstorm. General weather and its relationship to the operation can be just as strong a factor. There were three air carrier accidents in a rather short period. The first found a first officer making a takeoff in bad weather and the airplane crashing with fatal results. Possible diversions were factors in the second, and the possibility of ice on the wings was raised in the third. The 737 at LaGuardia and a Metro at Raleigh–Durham could have involved diversions, problems with systems on the airplanes. There was talk of possible ice on the wings of the DC-9 because of a lengthy wait for takeoff in a snowstorm. The two bigger jets stopped more or less on the airport; the Metro lifted off, climbed, apparently overbanked in a

turn to the right, and crashed. Although it is not likely that weather had everything to do with any of the three accidents, none of them, or perhaps only one of them, would have occurred on a clear day. Take the clouds and rain or snow away and it is a lot easier to manage diversions as well as risks. That is why it is so important we consider every factor before every takeoff. I don't ride the airlines much, but when I do I would hope that the captain has weighed all the factors, including the experience of his right seat mate, before deciding who will make the takeoff in grungy weather. Take poor weather, add a pilot flying without much experience, add a diversion, and you have the very real possibility of departure problems.

Not Always As You Think

Although we can usually make a pretty good pre-takeoff nowcast on the conditions that will be encountered right after takeoff, it isn't always possible. At the times when an incorrect assessment is made, we still have to manage the situation.

One late-winter morning in Wichita, I was planning a trip to Little Rock. There was dense fog at the airport so the first decision had to be about making a low-visibility takeoff. There is risk here, but I would be flying out over flat and relatively unpopulated country and was alone. So I rationalized that the risk would be mine and that it was slight. The next chore was to put together a mental picture of the expected conditions. I had nothing from the flight service station to indicate to the contrary, so I felt that the departure would be like many made before in an upslope condition in the plains. The expectation would be to take off and climb in smooth air to on top a few thousand feet above the ground, then enjoy a flight above a carpet of clouds and bright blue sky everywhere else.

That was the plan as the takeoff started—track the runway stripe and accelerate to 10 knots above the normal lift-off speed, then transition to instruments and fly away. At a couple of hundred feet, though, the script changed. The airplane flew into some light turbulence and the rate of climb and airspeed increased markedly as the airplane ascended. The surface wind had been calm. The good news was that I had taken off toward the south and was climbing into an increasing wind, thus increasing performance. Had I taken off north I would have been dealing with deteriorating performance, which is always more difficult. The bad news was that I would probably have a headwind. And the original script on enroute conditions was out the window. At cruise I was in bumpy and wet clouds and remained in them much of the way to Little Rock. The departure had been early in the morning so there were no pilot reports available on tops. As the system woke up, with more airplanes out flying, I found that the tops were well above the capability of the Cherokee Six I was flying at the time. The sign that hadn't been anticipated—the strong low-level southerly flow—was a clue that the rest of the morning wouldn't be as planned. Hunker down and fly.

Bumps and Temps

Turbulence and temperatures can tell you a lot about the way a day of cloud flying might work out and can give a quick grade on a forecast. If, for example, we take off with the thought that there will be stratus clouds out there but the air is turbulent, perhaps the forecast that led us to think it will be a smooth flight is incorrect. If a look at the outside air temperature gauge shows it to be substantially different from the projected temperature on the winds aloft forecast, then there is even more reason to be suspicious.

If this happens, the next step is to determine the nature of the turbulence. Is it wind shear or convective turbulence? The difference between them may not seem obvious at first, but with wind shear turbulence the airplane bounces around and there are airspeed excursions but not a lot of vertical motion. Flying through cumulus we get good jolts plus up- and downdrafts. It's a lot easier to tell the difference when you are flying level because if climbing, for example, into a rapidly increasing headwind, the big spike in rate of climb that would come as you maintain a normal climb airspeed might appear as vertical motion.

In flying the lower altitudes, below, say, 12,000 feet, we can usually climb above low-altitude wind shear turbulence. If flying higher, in the flight levels, the turbulence is associated with the jet stream or jet cores within the jet stream. This turbulence often starts from 15,000 to 20,000 feet, intensifies as you climb, and then starts to abate when you get above the stream or core. It might take as much as 45,000 feet in some cases. If the turbulence is convective, the upper limit of it may or may not be within reach of a normally-aspirated airplane.

Add the temperature to this for another piece of the puzzle. If the temperature is warmer than forecast, the wind aloft will likely be more southerly than forecast. This may also be linked to bumps because if it is warmer in the lower altitudes than was predicted and as cool as forecast at higher altitudes, then the air is more unstable than was anticipated.

The Morning Factor

I've always been a morning person and people who fly with me often joke that I must eat breakfast at their normal bedtime to make those dawn patrols, especially in the summertime. Early flying does make weather more interesting because the observations made at night are more suspect and fewer. There are also fewer airplanes flying, especially in the lower levels, and pilot reports are not as complete at night. You can shine a light on ice, feel bumps, and see lightning or the effect of wind shear, but that is about it. No judgment can be made, for example, that it is dark to the west or that tops appear to slope upward to the west. So although you might fly with almost as much information at dawn as for a noon departure, the information just isn't quite as good. This is where reading the signs becomes even more important.

Fog is a good example of this. More than once I have driven to the airport with a forecast that did not call for fog. And the stars would be up there twinkling away at my drive to the airport. But wisps and halos adorning streetlights and starting to show in the car lights give a clue. The airport might have that foggy feeling, with the hangar floodlights starting to show fog. By the time the airplane is out and the sun is just starting to come up, the visibility might be down to one in fog. Read the signs. This means you have to accept a low-visibility takeoff to fly. A fog that forms like that, just at sunrise, tends to dissipate rather quickly, so a little patience is all it takes if you don't want to charge down the runway with low visibility. Even in a shallow fog the low-visibility takeoff has the same requirements and risks that are found where it is foggy at the surface and the tops are 10,000 feet.

On a trip to Ohio from Maryland, departing soon after sunrise, I got another example of how this works. The ceiling was reported as 12,000 feet or above for the first half of the trip. File for 8,000, stay beneath the clouds. Piece of cake. Actually, the airplane was in cloud at 6,000 feet in the climb and remained in cloud for over an hour. That was not my expectation of how the first hours of that morning would work, but you have to take what comes along and not let an unforecast condition become a substantial diversion.

Diversions

Diversions are a bad thing at any time when flying, but especially so in the first few minutes of a flight. We have to put special effort into learning how to handle them. Unexpected weather can be a substantial diversion, too, because it can't be thought through in advance. If you get caught short on anticipation you can make up for it only by carefully tending to the basics. There is a tendency to feel betrayed by the weather dissemination system when conditions are unpleasant after they have been forecast pleasant, but this isn't a good idea. Just feeling betrayed is a diversion in itself.

I found an interesting diversion in Wichita one winter day. We were leaving about midday, and four or five inches of snow had fallen that morning and the snow was still on the ground. Because the ramp and taxiways appeared to have been plowed, I thought the runway would be clear. The fact that the air carriers were flying strengthened this thought. When we got to the runway, though, it was snow-covered. For the take-off roll I tried to follow tracks of other airplanes, but was only marginally successful. The airplane seemed to accelerate a bit and then slowed down. Finally, when the airspeed got to 60 knots, I increased the flaps setting and the airplane flew off into the ground cushion. Then came the awk-ward business of getting the airplane configured for a normal climb— the sinking spell with flaps retraction and all the sensations of accelera-tion. While the airplane was flying in snow, not in cloud, there was still no forward visibility. As we climbed, the northeast wind that had been at the surface increased dramatically for about the first 2,000 feet of climb, resulting in climb performance much above normal. Weird feelings in there. Keep the power at normal, the nose up about 8 degrees and the wings level, and everything works well, though. The combination of the abnormal takeoff and transition to climb and unexpected spike in climb performance from the increasing headwind had been weather-induced diversions not considered in the original plan. Anytime we come to a condition where the question in mind becomes, "Whoa, what's going on here?" best concentrate on the basics.

Ice

Ice is not easy to deal with in light airplanes. It is almost always in the forecast during the wintertime, and even pilots with deicing equipment have to make a plan that will work if the first signs after takeoff indicate that ice might be a problem. Deicing equipment is no cure-all; there are cases on record of aircraft with deicing simply not being able to stay aloft in severe icing conditions because of ice accumulation on unprotected surfaces.

There are many cases on record of airplanes getting into ice trouble soon after takeoff. Most of these could have been avoided had the signs been heeded, either before or immediately after takeoff.

One winter day we were planning a trip from Maryland to Florida in my deiced 210. A friend was planning the same trip in a Piper Arrow. Ice was forecast. The weather was not good, with rather heavy rain falling into a surface temperature of 1°C. An inversion really wasn't forecast, but with a warm front just to the south and heavy rain falling in Maryland, there was surely warm air aloft. The weather at the airports I consider to be takeoff alternates was good enough and there were no overnight pilot reports of ice anywhere. I opted to go. The Arrow pilot stayed on the ground—wisely. I had more answers to any questions that might arise. I would add that if there had been any pilot reports of icing, I would not have flown that morning. The possibility existed, given 1°C on the ground and heavy rain, that there could have been a layer of freezing rain 1,000 or so feet thick beneath the warmer air, and with the heavy rainfall rate it isn't likely that the airplane would have been able to climb through it. I have never seen that happen but it probably can happen, and is likely the sort of thing that you get to see only once.

More on Temps and Bumps

Watching the temperature in the beginning is an important part of reading the signs, especially if there is the possibility of ice. A general assessment of the conditions encountered is also a good idea. On the flight just mentioned, I looked at the temperature at about 1,000 feet on the climb and saw it to be still on 1°C. On my 210, ice starts forming first on the wings out near the tips when the temperature is barely at freezing, and a glance out there showed no ice. There were some bumps in the clouds as I climbed, but this would be expected when ascending into the slope of a warm front. Soon after the first bumps, and accompanied by an increase in the rate of climb as the airplane adjusted to the southwesterly flow aloft, the temperature jumped—enough for some fogging of the windows. It went up to about 7°C between 1,000 and 2,000 feet above the ground. On this one, everything went according to plan, but had it not I had an alternate plan.

The Airplane

The airplane is part of the overall system, and we have to be alert to any sign of trouble there. The first part of a flight is the most likely time to signal that something is not right, so it is important to keep your finger on its pulse from the very beginning. On takeoff, for example, when the power is set, you have the opportunity to make certain all the systems come up to speed along with the engine. You have a chance to glance at the exhaust gas or turbine inlet temperature to make certain it is on a normal reading for takeoff. And you have all the sounds and the feel of the airplane to judge as normal. If anything is out of place, you have the option to abort the takeoff early in the run. That would be between you and the airplane. Once off, it would be a return to land or a diversion to a takeoff alternate. If IFR in instrument meteorological conditions, that would be between you and the airplane and the weather and the air traffic control system. That makes it a much more complex operation, involving many more elements, and shows why it is so much better to abort a takeoff if there is doubt than to take those doubts for an airplane ride.

When sounds are mentioned, some pilots think about noise-attenuating headsets. There is a school of thought that these should not be used because you can't hear what is going on. This just isn't the case. In fact, if you don't wear them you may eventually not be able to hear what is going on anywhere, in or out of the airplane. I find that I hear the sounds of the airplane just as well with a headset as without, and the comfort and convenience of them is wonderful.

The takeoff alternate (which is not required for personal or business flying) serves two purposes. If you can't return to base, it gives you a place to go in the event of a problem with the airplane, as well as a place to go if the weather is so far from expected that an immediate retreat is called for. The latter shouldn't happen but it does, and when it does the key is in reading the signs and getting back on the ground.

I thought a lot more about the takeoff alternate business after moving to Maryland, where my original home base of Carroll County had only a VOR approach with high (754 feet AGL) minimums when I moved there. On many morning departures the option to return and land would not be available. My airport of choice in that case would be Dulles, which has acres of concrete and the finest in approach guidance, or Frederick, closer but with higher minimums and facilities that are not as good. In both cases the flight to the takeoff alternate would be over a relatively unpopulated area, which is where I would rather be in an airplane in need of help. A flight to Baltimore would be less desirable because it would be over a major urban area for a while. An important part of being able to handle something like this smoothly is in being prepared. Know the weather at the takeoff alternate and have the instrument approach charts for that airport at hand.

Since moving to Hagerstown, which has an ILS plus a VOR/GPS approach with a very low (292 feet AGL) minimum descent altitude, I haven't thought as much about takeoff alternates except on the rare occasions where a departure has been made with the weather below landing minimums for the airport.

I have never had to go to a takeoff alternate and should the need ever arise to go there it would be a real one and I'd sure convey it to the controller in a straightforward manner. If it were something like a door that came open, that would not require anything other than a normal sequencing for an approach. If it was something like smoke in the cabin or a rough-running engine, I would make sure the controller understood what was going on.

The Pilot

We have to read the personal signs before and during a departure because it is not possible to make a completely accurate assessment of how sharp we happen to be at every given moment. For an early-morning departure, we don't know ourselves very well for the day yet—at least the acquaintance has been a short one. I know in the flu season I have driven all the way to the airport only to decide, once getting there, that I really shouldn't go flying. The thought of punching into those low clouds just loses all its appeal. On the other hand, after being up all day you know how things are going and I don't think I have ever started to the airport to fly in the afternoon or evening with any doubt about feeling okay to fly.

This is especially important on a takeoff in instrument conditions or at night—which might as well be called instrument conditions unless there are a lot of lights around. We talked about the sensations of the initial phases of flight earlier. They are such that you sure wouldn't want to try to adjust to them if you're a little fuzzy in the head or otherwise not totally sharp. I recall reading some years ago about hangover studies done on pilots. One phenomenon that was noticed was a sort of fluttering of the eyes. I don't know whether or not you have ever experienced this, but it clearly wouldn't be conducive to good instrument flying.

Com Trouble

Another diversion that has to be managed carefully relates directly to air traffic control. Cloudy day, bumpy ride, contact departure control, no answer. Try again. No answer. Go back to the tower. No answer. Because we are used to flying IFR in constant communication, this is a strong diversion. We have to determine whether the avionics are not working as they should or we just have a wrong number. Just fly the clearance, they say, but clearances are often based on talking to the controller. Many contain a routing that includes radar vectors, and there is no way to get a vector if you can't talk to the controller. There can be a lot of things in mind, but the clearest message of all should be that the opportunity to solve all those other problems is based on keeping the airplane under control.

VFR Signs

The signs you read for VFR flying weather are more straightforward than the IFR signs. Some are the same, such as a wind stronger and more southerly than forecast, meaning the chance of weather worse than forecast. The clearest sign though, and the one we have to always heed to stay out of VFR trouble, is in being able to see as far as planned at the lowest safe altitude as determined before flying.

The valuable signs we get while flying certainly aren't limited to the departure, and much of a pilot's success in dealing with weather relates to the ability to evaluate what is seen and felt on a continuous basis and to do something about conditions that aren't anticipated. For some reason, though, I get the biggest kick out of that first discovery after take-off. It's sort of like jumping into the swimming pool when you don't know the temperature of the water. Maybe it's colder or warmer than you thought it would be, and it is that initial entry into the existing condition that defines something about how the swim will go. To relate this to flying, I had a friend who flew only for pleasure, usually in the local area. There was nothing adventuresome about him; he would take off and if the air didn't suit him perfectly—meaning it was bumpy— he'd scurry around the pattern, land, and put his airplane back in the hangar. Nothing wrong with that; it is just an example of how the flying can be more personally satisfying if you read the signs and land if they don't suit you.

Chapter 6
VFR Weather

Although VFR cross-country flying is thought of as the province of newer pilots, those who don't yet have an instrument rating, it is really for everyone. And it may start to become more popular again after years of giving ground to increased instrument flying. VFR flying is one of the more unique aspects of general aviation, offering almost complete freedom and independence and a lot of flexibility that is lost in IFR flying. Never saying hello or goodbye to anyone, we can crisscross the country, flying between any of the many uncontrolled airports. As IFR flying becomes ever more complex and demanding, and as IFR procedures result in more convoluted routings, more pilots might seek out the freedom of VFR flying. True, it takes more effort than ever to avoid regulated airspace when flying VFR, but that shouldn't come as a surprise to anyone. Twice as many people are flying the airlines today as ten years ago, but fewer people are flying general aviation airplanes than ten years ago, and those who use the airspace more get more of it. That works for IFR, too. But there is still plenty of VFR airspace left. The real rub on going places VFR comes not from regulated airspace but from the weather. Unless it cooperates, we can't go; it is that simple. If pilots try to fly VFR when the weather is otherwise, the risk involved goes out of sight. VFR flying in marginal weather has proven to be, without a doubt, the highest risk form of flying.

VFR Not Recommended
The FAA, through its flight service stations, has long had a "VFR not recommended" policy; briefers would tack that warning onto a briefing

if the actual weather or terminal or area forecasts showed the possibility of below VFR conditions along the way. This was how the FAA approached the "Continued VFR" accidents that were plaguing general aviation. The intention was no doubt good, but many pilots felt that the policy resulted in too many cries of wolf. There has been some truth to this, perhaps not as much the fault of the FSS specialist as of the preparers of forecasts, who endeavor to cover every base as well as their derrière. Maybe this policy is going to go away, too, unless the FAA manages to tack the warning onto all the computerized briefings that we get at home, at the office, or at the FBO. What they have to realize is that pilots don't want to go out and ram a foggy hillside, but they do want good information. There is also a great need for better education on VFR cross-country flying when the weather is not so good. Currently there is little training material available on VFR in other than excellent weather. Most dual cross-countries in training are conducted in good weather. Virtually all solo cross-countries are dispatched on days when the weather is excellent. Pilots are pretty much left on their own to determine how to make the right decisions on days when VFR becomes dangerous. The record shows that this isn't currently done very well. This applies to some old-timers as well as new pilots.

Average Wreck

VFR accidents involve a cross section of pilots, but the NTSB was once able to define an average VFR weather accident, pilot, and situation. The pilot had a private certificate and between 100 and 300 flying hours. His age was between 41 and 45, and one passenger was along for the ride. The purpose of the flight was personal business or pleasure. It goes without saying that the end came in IFR conditions, likely in rain or fog. The NTSB study didn't cover the condition of light, but these accidents often happen at night. In fact, in one compilation of VFR weather accidents, almost half of them occurred in the dark. At the time there was no difference in day and night VFR weather minimums. Although this defines a problem, it does nothing to suggest how we might better make judgments of VFR weather.

CVFR

When I started flying in 1951 it was VFR or nothing. Few general aviation airplanes were equipped for IFR. And most good instructors would take students flying cross-country on marginal days, in an effort to teach them how to survive out there in that misty world. I had one instructor who actually took me on a scud run to show me why it didn't work. It was a convincing demonstration that ended with a pull-up into clouds, a 180, and a descent back to visual conditions. That was probably not a wise thing to show a student because it might make him think it an okay procedure but, in my case, it was convincing.

I revisited scud running in 1975 while writing a series of articles on "Continued VFR" for *Flying* magazine. The object was to actually show, in black-and-white pictures, what it looks like when you fly to the end of the rope in VFR conditions. I had the blessing of the FAA's Southwest Regional Office, through regional administrator Hank Newman, so I got the complete cooperation of the air traffic control folks. Famous aviation photographer Russell Munson did the original photo work. For one series, the deal was for the ATC folks to block the minimum enroute altitude of an airway that roughly followed the route of an interstate highway. We would fly down the highway, photographing and maintaining VFR until that was no longer possible. Then we would pull up to the MEA, give ATC a position, and they would clear us back to the airport for an ILS approach. Then they would give us a Special VFR out of the control zone and we'd do it again. That was a little unorthodox, but I knew the territory and felt there was little risk.

Classic Pictures

One day we were able to get an absolutely classic series of pictures of continuing VFR down a road, flying at about 500 feet, until conditions completely went to pot. We used three of the pictures in the magazine. The first showed light rain on the windshield, with the highway plainly visible. In the second picture the rain was harder and the road barely visible over the nose of the airplane. In the final picture the rain was hard and the road but a blur in the lower left corner of the windshield. In other photographs in the same series we showed a road disappearing into an obscured mountain pass, a between-layers situation that finally merged, and a TV tower poking up out of the clouds. The FAA liked this series so much they asked us to do a movie. Alas, though, doing things like this through the Southwest Region and through Washington bore no similarity; with the conditions that Washington wanted to put on the operation it simply would not have been possible to take the pictures.

Review

It was an interesting project, one that reminded me of the exigencies of VFR flying in other than perfect weather. By evaluating weather and actually looking for impossible VFR situations to photograph, we learned a lot about what works and what doesn't.

In almost every case, if we could find general rain we could continue VFR until it was no longer possible. Although there are some rain conditions in which the weather remains VFR for a while—rain falling from higher clouds north of a warm front will usually do this—a lower layer of scud eventually beings to form. Then those low clouds start to multiply like rabbits and, presto, you have unflyable VFR weather. The scud that forms in rain often seems to hang in the treetops, too. Rain simply means that VFR won't be possible sooner or later. Information on rain is available from the radar reports and the radar summary chart. Another good piece of information for the pilot flying VFR is the weather depiction chart, which shows areas of IFR and of marginal VFR, defined as a ceiling between 1,000 and 3,000 feet and/or visibility between three and five miles.

Remembering all this business about rain served me well on the first day of the air traffic controllers' strike in 1981. I managed to get an IFR clearance as far as Cleveland that day, en route to Oshkosh for the air show. At Cleveland though, the weather and my clearance went sour. We landed, hung around the airport for a while, and watched President Reagan on TV read the riot act to the controllers. Then it was VFR time, and we were soon rocketing down a highway, reading town names off water tanks and fun stuff like that, while committed to a retreat if not able to fly 1,000 feet above the ground with three miles visibility. All was well save for one thing. There was an area of moderate rain ahead, as seen on my 210's weather radar. With the weather at my minimums for VFR I knew rain would make it worse, so I landed and waited a while. Soon I was back at it again, and it was possible to get to Wisconsin that day without breaking any of the rules of running the scud that served me well in days gone by.

Stormy Weather

There is even a time when the weather isn't good it is better to fly VFR than IFR if you fly at all. I used to run with this a lot when I lived in the south in relation to thunderstorms. There are simply times when you don't want to be where you can't see, especially if you don't have a Stormscope or airborne weather radar. If you can see the rain shafts and examine the appearance of the clouds, it is possible to make better decisions on whether or not the condition is flyable. This is especially true when severe weather is forecast. When living in Arkansas I saw two tornadoes from my airplane—photographed one—that were in the process of dismantling small towns, yet I was a safe distance away and enjoying a good ride. In one case a jetliner crew reported turbulence in the clouds up higher, and when I told them it was smooth down low, they joined me beneath the clouds. There are also days when there is ice in the clouds and none beneath, making VFR the more attractive alternative if the ceiling and visibility allow.

Another time when it is more comfortable to be VFR down low is when there is a trough aloft, as shown by those dark-bottomed mammatus-looking clouds, sometimes with streaks of virga (rain that doesn't reach the ground) beneath them. The air can be quite turbulent several thousand feet below the bases of those clouds and be reasonably smooth down lower but still 2,000 or 3,000 feet above the ground.

Some IFR pilots choose VFR when westbound into a strong wind. It is usually true that the wind is lighter down low, but it is also true that it is usually more turbulent as well. Personally, I usually climb to smooth air in such a condition.

Air Rules

The difficult part of getting some measure of utility out of an airplane flying VFR comes in drawing the line on what is possible and what is not, what is risky and what involves low risk. The regulations don't really do us much good in helping us make these VFR judgments. Whereas IFR flying is procedural—match the numbers on the instrument panel with those on the chart—and is not based largely on the pilot interpreting the view out the windshield, VFR flying is all based on individual interpretation. No forecast or report counts for as much as what we see and what we decide to do about what we see. To a VFR pilot, instrument flying may look difficult, but of the two it is actually the easier.

Perhaps the most important thing a VFR pilot keeps in mind is the general weather synopsis. Doing this gives a sense of where conditions are likely to be better and where conditions are likely to be worse. I recall one windy spring day starting off VFR in my Skyhawk, headed from Little Rock, Arkansas, to Kerrville, Texas, and running into more questions than answers. There was a low to the northwest and a cold front to the west of my proposed route of flight. The ceiling was a couple of thousand feet and the visibility wasn't bad, but the air was very rough.

The tops were higher than the Skyhawk would go, which was one reason for VFR. I decided that the turbulence would be worse in those clouds and the headwind would be even stronger up there. Even though I was flying parallel to the front, it was moving toward the southeast, in my direction, and I knew that if there were to be a change in the weather, it would get worse instead of better.

Two things made me abandon this trip. One was a convective SIGMET telling of a line of storms that was to the west, but that would move across my route while I was still trying to get to Kerrville. Persistence might have made me keep it up, to see if I could find a break in the line or land somewhere and let the line pass by. But the other conditions, the ceiling that was getting a little lower and the visibility that was getting a bit worse, said no. I didn't want to go look at a line of storms that I could not see from a fairly substantial distance. Also, I called a center controller and asked how the line looked. His one-word answer, "Solid," helped make the sale on a 180 and a retreat back to home base.

On that flight, the briefing I got before leaving suggested I could get to Kerrville before the storms. What I saw, and what I learned after takeoff were far more valuable than the preflight briefing in making the decision not to continue.

Cold Front

The possibility of getting through a cold front VFR is not always good, but it is more likely than getting through a warm or stationary front VFR. The nature of warm and stationary frontal weather—low ceilings and rain and poor visibility—is just not what VFR is made of. And a frequently made fatal VFR mistake is continuing toward a stationary or warm front, or toward a low-pressure storm system, with the feeling that it will be possible to fly through. The simple fact is, when flying toward such a system, the weather can only get worse. If you are flying beneath the clouds, they will get lower. If you are between layers, with scattered clouds beneath, they will only thicken and the layers will eventually merge. I went out one day in a pre-cold frontal condition to do some videotaping in a Piper Cadet. I had just landed in my 210; the southerly flow was strong and the air quite rough below about 6,000 feet. The weather was 5,000 scattered and 25,000 broken. Within 30 minutes we were up in smooth air in the Cadet. But within another 20 minutes the scattered clouds at 5,000 feet became so numerous I had to do a power-off spiral to beneath the clouds in order to maintain all the VFR cloud clearance requirements. It happens that fast, and a pilot who must finish a flight VFR (as opposed to air-filing and getting an IFR clearance) is out looking for trouble flying on top of any clouds other than widely scattered ones. This is true anywhere and even more true if flying toward any sort of front or low-pressure system, or if flying in an area with a front or low approaching the area. One thing that has always been true is that one or more airplanes are likely to be lost whenever a large weather system moves across the country, especially on a holiday weekend. Some just try too hard to get there VFR.

Cumulus

Summer days pose a VFR weather challenge because it is much more comfortable above the bases of the cumulus that form most days. Usually the morning works fine, with the little puffy cu starting to form about midmorning. But they multiply as the heat builds and they also build upward, with some aspiring to be one of those scattered afternoon thunderstorms. It is hard to say when a VFR pilot should give up on the cool and comfortable flight above the level where cumulus start to form. The swap for the hot thermal turbulence beneath the clouds is simply not a pleasant choice. But there is a proven track record of pilots staying high too long and having a problem getting down or, worse, winding up in a thunderstorm. On average, as many VFR as IFR airplanes are lost to thunderstorms. This might happen in a few different ways. One is for a VFR pilot to stick to a higher altitude as the cumulus build and become more congested, only to wind up in a blind alley with a thunderstorm at the end of it. Certainly it takes a lot of space to do a 180-degree turn, and even a cumulus with tops of 10,000 or 15,000 feet would be mean to a VFR pilot who entered the side of it in a fairly steep turn while trying to beat a hasty retreat. If the cu had reached the level of free convection and was building into a thunderstorm, it would be even less friendly.

Another way to fall victim might be in getting caught fudging—that is, busting through ever-increasing cumulus until reaching one that had achieved a complete state of development. Still another way would be to try to fly just underneath the base of a developing thunderstorm at about the time it matures. One might imagine that you could be sucked up into the thunderstorm clouds and I suppose that could happen. The updraft, though, is generally around the outside of the mature storm, with the downdraft in the center. One thing that is certain is incredible turbulence as you fly from the air affected by the inflow into the storm, then into the air under the influence of the outflow. The flight recorder of the Delta L-1011 that crashed in a storm at Dallas/Fort Worth showed that even this large airplane was subjected to almost unmanageable turbulence as it passed through this storm, probably below the actual base of the storm, as it was in the last throes of an ILS approach.

Still another way to get caught might be on a hazy summer day, with thunderstorms lurking in the haze. If visibility is pushed to the minimum you can literally find yourself face-to-face with the side of a thunderstorm you didn't suspect was there a few moments ago. Some pilots say that you can avoid thunderstorms buried in haze by flying along looking upward at about a 45-degree angle, searching for dark and light spots and, naturally, always flying toward the light spots. Sounds like a pretty desperate measure to me, though.

Summer cumulus have to be dealt with by swapping the cool and smooth air for the bumpy air with a view beneath the bases of the clouds whenever either dodging the clouds or the distance available between clouds for a letdown raises the slightest question. The ultimate bad scene would come with a VFR pilot flying a turbocharged airplane winding up at, for example, 17,500 feet with cumulus building through that level and becoming congested. The hazy thunderstorms of summer can be handled VFR only by flying when the visibility is good enough for early sightings and by staying beneath the bases of the cumulus when the clouds become numerous or tall.

Night

We don't do well at all flying VFR cross-country at night—the record is terrible. Actually, VFR at night in marginal weather is a contradiction in terms: If the weather is marginal at night, it simply is not VFR.

The problem at night is in not being able to see. In addition, not as many weather observations are taken at night, which compounds the problem. This makes the continuous evaluation of weather during night VFR cross-country quite critical.

The best advice to follow is to do it only on clear nights with no mention of ground fog in any of the forecasts and with wide temperature and dew point spreads. Then the requirement is to be meticulous with navigation, even on a night with good weather. There's terrain out there you can't see, and more than one pilot has simply flown into terrain because he didn't know it was there.

Some classic VFR night accidents occurred in Arkansas when I lived there in the '70s and they have continued to follow the same pattern since. One reason for this is a widespread belief among pilots that

Arkansas is flat. One aviation writer even wrote once that Arkansas is for a fact flat. Nobody ever told the mountains that occupy the northwest quarter of the state that they don't exist, and these mountains have been assaulted by a wide variety of VFR airplanes, including military and airline aircraft. The number of airplanes that have crashed in these mountains has been bandied about for years—I have heard as many as 60. The scenario is usually the same. The weather is bad, it is often night, and the pilot probably doesn't realize that although the elevation of the peaks is nothing like the Rockies, they do start from close to sea level and extend up well above 2,000 feet. In one accident in the early '70s, documented with a cockpit voice recorder because it was an airliner, the crew was attempting to deal with a line of thunderstorms at night VFR. They were far north of course in their attempt to find a gap in the line. In fact, they were into the mountains of northwest Arkansas. Not long before the aircraft impaled the side of a mountain, the captain said something to the effect that he knew the highest terrain and it was well beneath them. There is sort of a double jeopardy in connection with these mountains, too. In the winter and spring, fronts tend to become stationary in the area, adding a touch of temptation for those who profess to being able to get there by running the scud.

GPS

We have some fine electronic VFR help in many of the new GPS sets. Among other things, many will give minimum safe altitudes for the present position as well as for the route selected in a flight plan. If you use one of these, you'll quickly see that the minimum safe altitude is seldom very low. Even in flat country it can actually be very high. Taken to the extreme, flying south along Florida's east coast one day I dialed up the minimum safe altitude on a King KLN 89B GPS and it responded with 17,700 feet. You hardly have to fly that high to be safe if you know where you are, but there was a tethered radar balloon south of Cape Canaveral that extended up to 15,000 feet or higher. This is an extreme example, but there are a lot of man-made obstructions out there along with terrain such as that found in northwest Arkansas. And if you don't know precisely where you are in relation to obstructions or terrain, it is the highest obstacle that counts. You could use this GPS feature in determining the minimum acceptable altitude for a VFR flight. Another fine GPS feature for the VFR pilot is the "nearest airport" capability. At the press of a button you can know the distance and bearing of airports that are close by. Runway and communications information is usually there so if the going gets murky, information on alternatives is available. Real men and women can do all this with a map, pencil, plotter and Waterbury watch, but isn't it nice to have a little help?

Coming online in the early 2000s, terrain awareness and warning systems that play through a multifunction display will tell all about the terrain in relation to the height of the airplane, using GPS position, and moving maps provide excellent information on the whereabouts of the airplane in relation to points on the ground.

GPS is so good and useful that were I equipping an airplane for VFR flying, as opposed to IFR flying, I would put a good GPS into the panel right after a communications radio and a transponder with Mode C. An alternative would be one of the fine combination GPS nav/com units and a transponder. GPS is really all you need in the way of VFR navigation information and a low-cost handheld unit can be an adequate backup for the installed gear.

Whether flying day or night VFR the key to success is in making a plan, following it, and always knowing where you are. If, for example, you can't maintain VFR at 2,000 feet, then it is time to go somewhere and land—no exceptions. The pilots who get in trouble VFR are the ones who press on because they think they can make it.

Judging Visibility

Judging inflight visibility is not difficult and it is another key to staying out of VFR trouble. This is especially critical on those hazy summer days when thunderstorms become embedded, as well as when making a judgment on whether or not the visibility is good enough to continue in any weather condition.

One way to judge flight visibility is to refresh yourself on 30-60-90 right triangles. You have only to remember that in such triangles the longest side is twice the length of the shortest side. The short side is from your eye to the ground. The long side would be 60 degrees up from that vertical line. If flying at 5,500 feet and looking out the side of the airplane down an imaginary line 60 degrees up from a vertical line beneath you, you'd be looking at a point about two miles away. If you could see clearly to a point twice as far away as that, the inflight visibility would be four miles. Another way to judge visibility takes a little groundwork with a specific airplane. You need to know only two things—your eye height above the ground with the airplane at rest and sitting level in a fore-and-aft sense, and the distance from a point on the ground directly beneath your eyes to the closest point you can see over the nose of the airplane. On one I measured, my eye height was five feet above the ground, and the closest point I could see over the nose, in a normal sitting position, was 35 feet. That's a ratio of seven to one. Were I flying along at 2,500 feet and looking at a point over the nose, that point would thus be 17,500 feet away, or about three miles. In that airplane I'd know that at 2,500 feet, being able to barely see points clearly over the nose would mean that the flight visibility was at the minimum.

Visibility is an area where the regulations desert us in making determinations of what is safe and what is not. A blanket rule that sets the VFR minimums in miles does not take into account differences in speed. What really counts on visibility is time. In a J-3 Cub, three miles visibility is three minutes—plenty of reaction time. In a Bonanza, three miles is scarcely over one minute—not time to do much, especially if you consider a much larger turning radius. The faster the airplane goes, the less time you have to do something that takes more space—not a good deal. When considering the one-mile visibility that is legal in Class G airspace, a Bonanza pilot has less than 30 seconds to evaluate conditions ahead and react.

Jack Poage, an extraordinary aviator and my FBO at Carroll, and I were standing on the ramp one day watching as a King Air pilot tried to pull off the scud run part of the nonprecision approach to the airport. Jack observed that if he were out there doing it he might be able to land—but only in his Husky flying the approach at 60 knots.

Not All Clouds

Not all weather is in cloud form. Wind can be very much a factor in flying, and forecasts of wind are not always accurate to the point that we can definitely say, for example, that the wind at Indianapolis won't be in excess of what I consider a maximum when I arrive there in three hours. The surface wind limit is strictly up to the individual pilot, but it is a number that we should all have in mind when we fly. At some point the surface wind velocity exceeds the pilot's ability, or the airplane's ability to remain on its feet.

I recall a fall day years ago when the wind almost got the best of the Piper Pacer I owned at the time. The goal was to fly from Linden, New Jersey to Cleveland, Ohio. There was a cold front just to the west of Cleveland, moving east rather rapidly. There was little weather in the front, and from what I could learn the surface winds were not over 30 knots, which was the velocity where I started becoming very wary in the

Pacer. A taildragger has a more finite wind limit for taxiing than a tri-cycle mainly because it is always at a positive angle of attack when all its wheels are on the ground.

After leaving Linden I started getting plenty of clues about wind. The ground speed was far lower than I had anticipated, and as I sought weather information along the way, the surface wind was much fresher than had been forecast. The good news was that the weather was okay even though the wind was blowing.

The turbulence increased markedly at any altitude I could reach in the Pacer and, in general, the whole thing was becoming quite uncom-fortable. I had a bit more to learn about weather at that time, and my first solution to the challenge was to land and check weather further. Trouble was, I picked the worst possible place to do this—Williamsport, Penn-sylvania. The airport there is down in a valley with a ridge to the south of the airport. The strong southerly flow was spilling over that hill and the reported gusts at Williamsport were probably higher than anywhere else in the area. The wind was splitting the difference between Runways 15 and 27, favoring 15. I made the approach to that runway but could see that landing was going to be an uncertain affair. The airplane was jumping around a lot and I simply wasn't satisfied that it would be pos-sible to land it with no drift—a definite requirement in a taildragger. The wind spilling over that ridge was just too bothered when it got to the runway.

So it had to be Plan B. I remembered flying over a nice wide grass strip at Bloomsburg, Pennsylvania, some miles back. Its alignment was about 260 degrees, still quite crosswind, but the width of the strip and the fact that it was grass did a lot to recommend it. Also, though I probably didn't consider this at the time, it was a little farther from the surface position of the front, which would mean the wind should be slightly lighter. Finally, there was no ridge to the south to contribute to the gusty and disturbed nature of the air. The landing there worked fine, though it was plain it had to be the last one of the day. The low-pressure area had strengthened and the surface wind was truly howling through the area.

Inflight Plans

We do have to be willing to change the plan en route if the surface wind goes beyond what appears to be a limit. It is generally true that the wind will be stronger the closer you get to a frontal zone, so this can be used as a guide when seeking information on airports that might have lighter winds. The flatter the terrain, the steadier the wind might be. The turbulence should be less, too. The runway alignment at the airport is important, as are any obstructions around the runway. A tree line upwind of the runway, or buildings, might mitigate the wind a bit but will also make it shiftier. The better deal is to have a relatively clear area upwind of the runway.

In the Mountains

Wind in mountainous terrain can create more turbulence than we care to deal with or, in extreme cases, more turbulence than the old airframe can bear. Downdrafts in excess of the airplane's ability to climb can develop quickly. Anytime there is a strong flow in the mountains, anything near the 30 knots at ridge level that is considered a limit by many, best have alternate plans if the going gets too rough. We are dealing with a forecast of wind here, and forecasts are always subject to errors.

Flying in the mountains when it is windy requires a definite plan. Certainly this is one place where VFR is far better than IFR unless the airplane is turbocharged and can be flown at least 10,000 feet above the ridges, where the downdrafts can be better managed but where the turbulence can still be significant. Why is VFR better? Because IFR you follow the airways, or fly direct in some cases. VFR you can adjust the route to avoid the areas where the downdrafts and the turbulence will be strongest. I'll always remember one mountain-wise pilot telling me that he didn't even use turbocharged airplanes because one, they won't really go high enough to do a lot of good in the Rockies; two, they aren't as reliable as airplanes with normally-aspirated engines; and, three, success in flying in the big mountains comes from staying VFR, knowing where you are, and flying where you know the wind and the turbulence will be less of a factor. And the way to do this is to carefully visualize the flow of air over the high terrain. The absolute worst turbulence is found in a rotor that forms downwind of a substantial ridge. The top of the rotor would be at or slightly below the ridge level. The bottom of the rotor would depend on a lot of things, including the height of the ridge and the strength of the flow. But the message would be to beware the area downwind of a ridge. If flying in a valley, for example, you'd want to stay on the far side of the valley.

Passes

Mountain passes can be desperately turbulent places when the wind is strong because, as it squeezes through the pass, the wind can accelerate to a velocity far in excess of the general wind. Not being an inhabitant of the big mountains, I usually seek and heed the advice of the locals when making a decision on whether or not the wind suggests waiting until later to fly.

Apple Pie and Motherhood

There is a lot of VFR weather out there, and flying VFR is not only a great freedom, it is fun. Even the basics are enjoyable. When working on the "What You Should Know" video series for the private certificate with Sporty's, I got a good review of this and it sparked a lot of memories. I actually found an old sectional chart from the early fifties, one I had used on cross-country training flights. The course had been carefully plotted from Camden, Arkansas, to Little Rock. The calculated course was 15 degrees and I had marked the landmarks, of which there were few, on the map. The straight line was over the "big woods" of south Arkansas—tall pine trees; unless you flew right over them you might miss a road or a small village. As there were no avionics in the airplane to use in orientation, confusion could set in. It was 86 miles, so in a Cessna 140 it took about an hour. I flew it many times and enjoyed most of the flights very much while I learned some lessons about VFR weather. One of them relates to the Saline River—not a major river but in flat country and with a wide swath of bottomland. If the weather wasn't good, I would fly out of the way and follow a highway. The highway crossed the Saline and often I could get only that far. Conditions over the river bottom were always worse than elsewhere—at times it might as well have been a front. Good challenging stuff, that VFR.

Chapter 7
IFR Weather

Whether we leave on an IFR flight with a complete computerized weather briefing or with notes made during a phone conversation with a flight service specialist, we have to keep a continuous tab on what is happening while we fly. It is somewhat different than VFR, where by the nature of the type of flying you should be able to see. VFR, you are dealing with ceilings and visibilities while you fly beneath the clouds, constantly examining the conditions. IFR, our consideration of weather is more abstract. When we are clear of clouds and can see, that counts for a lot, though it is usually not the ceiling and visibility we see. More likely the cloud tops are visible, or we are flying between layers of clouds. To learn of ceilings and visibilities we have to call and ask or, as many pilots do, listen to automatic weather broadcasts. By listening to the ATIS, ASOS and AWOS broadcasts ahead and to the sides, we can, while flying along, get a good idea of how the weather is unfolding. When we are flying in clouds and can't see, most of the information comes from the radio, though what we see and feel is still valuable.

The basics apply as much to IFR conditions as anything else. Smooth flying usually means stable conditions. Bumps in the clouds suggest unstable conditions; the same goes for big noisy raindrops on the windshield. A temperature aloft warmer than forecast is usually accompanied by a more southerly wind aloft component than forecast, often at a stronger velocity than forecast. If anything, the surface weather might be worse than forecast. In any event, whenever a wind forecast is substantially in error it is a mandate to question all the forecasts obtained before takeoff because the computer model on which everything was based proved to be in error. The real success in dealing with weather while on an IFR flight is in being suspicious. If a pilot flies both VFR and IFR, the differences have to be recognized. In VFR the weather is updated moment by moment as you peer out the windshield and side windows. Flying IFR in instrument meteorological conditions, the pilot has nothing to see, and word on the weather ahead has to be obtained from some other source. The surface weather at the destination and the alternate—where the only minimums for the flight are established—needs a constant check, especially if other things are not going as forecast.

The latest equipment will let you access weather information from the airplane for viewing on a multifunction display on the panel.

Not as Advertised

I was flying from Kansas City back to Maryland one early-April morning and found an interesting weather example, one that was perplexing at best but that followed an old rule of thumb. There had been a closed low aloft over New England for several days. When there is a closed low, sometimes called an upper level low, with a complete counterclockwise circulation in the upper air, it becomes more difficult to forecast what happens next. The reason is that it is difficult to forecast with much precision the formation of a closed low aloft. Ditto the breaking up of one. Surface lows tend to form and move slowly toward the closed low aloft, enjoying that upper-level support. They don't move out the way surface lows usually do. The result is that the weather and the movement of systems don't follow the usual script where there are troughs and ridges aloft. I was aware that this was what had been going on over the northeast, but I thought, as did the forecaster, that the low aloft would break up and the day of this trip would see a dramatic improvement.

The forecast I got on the phone from the FSS agreed. It called for post-cold frontal conditions in the Baltimore area—broken clouds, good visibility, and a fresh northwest wind—and winds aloft at 18,000 feet from the northwest, becoming stronger toward the east. Usually, with a condition like this, you would envision fairly clear weather to the Appalachian Mountains, where there would be instability, clouds and snow showers with tops in the 10,000- to 12,000-foot range, and pretty good weather east of the mountains, about as forecast for Baltimore that day.

At Flight Level 190 the going was fine. The wind was slightly more northerly than forecast and there was some light turbulence indicating a bit of wind shear. The temperature aloft was few degrees above standard. Nothing suggested that the weather to the east would be other than forecast.

The first clue came when it was necessary to climb to Flight Level 210 to stay on top of altocumulus clouds. That was not the typical postfrontal condition that had been expected. The Baltimore weather gave another clue. The ceiling there was 2,400 feet, it was raining, and the wind was out of the west-southwest instead of the northwest. All this was enough to convince me that the synopsis was different than had been expected. The upper low was still there and an active surface low was closer to where I was going than had been anticipated. Still, a 2,400-foot ceiling isn't a bad deal. Then I heard that Hagerstown, not far west of my destination at Carroll County, Maryland, was reporting 800 overcast with two miles visibility in rain, and with a surface temperature of 4°C.

The "I" Word

The descent started about 100 miles out and the airplane was rather quickly in clouds. The going was neither rough nor smooth to begin with, but there was some ice in the clouds. Because it was to be a continuous descent I didn't worry much about this and actually cycled the deice boots only once. I could tell, though, that there was ice in the lower levels. The surface temperature at Baltimore was 4°C and pilots were keeping the controllers busy with requests for altitude changes. This was in the time when the FAA had become aggressive about "flight into known icing conditions," so nobody would use the magic word. They would complain only of turbulence, which was there too, and request a new altitude. The tip-off that it was ice was that pilots were asking for a lower altitude where the turbulence would likely be worse but the ice possibly better. I reflected that it is not good to fly in a system where being candid about your needs can result in legal problems. Nobody but nobody ever said "ice." Since that time the FAA has seemed to change its stripes—when a pilot requests a change because of ice they go out of their way to handle the request. This was prompted by two regional airline turboprop wrecks, plus some incidents, related to airframe icing.

Finale

The air in the lower altitudes was quite turbulent and the lower I flew the harder it snowed and then rained. This was clearly no postfrontal condition, and when maneuvering for the approach I knew there would be a lot of wind shear. There was enough for 20-knot fluctuations in airspeed. The weather at Carroll County Airport was close to minimums, with what felt like a half gale blowing across the runway.

It was an interesting bit of weather because of the amount of en route attention it took to realize that the conditions would not be as forecast for the arrival. More likely in IFR conditions we are dealing with southerly flows and troughs and ridges aloft. The only rule that worked that day was to beware all the forecasts when there is a closed low aloft. The prediction of it moving out did come true later, with rapid improvement in weather at home and a beautiful sunset. They were just off on the timing.

Overload

That flight was with a phone briefing from a specialist who was good at reducing the area forecast to some understandable information. That the total forecast didn't unfold as planned was not his fault. When we leave with a computerized briefing we have much more information but it can be as suspect.

This next one was an outbound flight from Maryland. The area forecast basically said that over the eastern United States there would be clouds to the moon, turbulence, ice, and everything else that you might imagine. Remember, the area forecast covers everything that might happen all day. What we have to do is look at what is happening now to make the decision. That morning I backed up the computerized briefing with the morning TV-show radar reports, which showed precipitation only well to the west, and the satellite picture which indicated that the murky appearance out the window of our house was fog, probably not too thick, and that any higher clouds above were not likely.

On the drive to the airport, the visibility ranged from a half-mile to three miles. At the airport it was foggy, but the visibility easily met my personal requirement for a low-visibility takeoff. That minimum is a highly individual matter, with no minimum prescribed by law in not-for-hire operations.

After takeoff, the airplane was on top of all clouds at 2,500 feet with clear skies above. There were breaks in the fog over the Shenandoah Valley. As we proceeded westbound, there were clouds ahead and we were soon in them. Some precipitation appeared on the radar and a few deviations took care of most of the showers. There were light bumps through the area, it was soon snowing, and then the weather improved and it was fine on to Kansas City. Instead of the eight pages of contractions in the computerized briefing, it could have been summed up with: "Shallow fog along the east coast, a cold front on a Pittsburgh-to-Cincinnati line with about 200 miles of inclement weather, improving to the west." The bothersome thing that day was that the forecast had maximized the possibility of poor weather over a large area and actually minimized the weather in the frontal zone. There it was worse than forecast for VFR flying. Several VFR aircraft were out there, you could hope they got a briefing; they were clearly finding worse weather than they had anticipated.

For Better or for Worse

In the preceding two examples we saw weather that was worse than forecast and weather that was better than forecast, at least for some pilots. Both were rather simple cases, related primarily to weather. In neither was there a strong requirement for an interface between the weather and air traffic control systems. I required a climb of 2,000 feet on one flight and some simple deviations around showers on the other, but had neither been available there wouldn't have been a lot of additional risk involved. This, though, is not always true, and we have to recognize that the air traffic controller's number-one priority is separating aircraft from each other.

Thunder vs. the Controller

Some pilots fly with a feeling that it is the responsibility of the air traffic controller to tell them of thunderstorms and provide whatever service is necessary to keep the aircraft out of these storms. This is not the case. If a pilot winds up backed into a corner with no apparent way out, it is the pilot's and not the controller's fault. Consider that we deal with air traffic controllers with varying levels of experience who may or may not understand or appreciate the difficulty that thunderstorms pose to pilots. Controllers used to deal with radar that is not weather radar, though they will have the ability to overlay high-quality weather radar information on their scopes sooner rather than later. This will help a lot. It will enable the controller to better anticipate what pilots are asking for. Having said all this, I still ask controllers what they see on their scopes when there is the possibility of thunderstorm activity. They might have a little bit of information that I don't have, and every little bit counts. Even more important is their overview of the big picture. A controller can often suggest a new route that bypasses everything on the scope. That is a better deal than diving into an area of precipitation and wondering what comes next. Uplinked weather radar information will help as more and more airplanes get the equipment.

Inflight Advisories

One thing I don't think we take full advantage of is the inflight weather advisory—in the case of thunderstorms, the convective SIGMET. This product is specifically designed to inform pilots of thunderstorms that exist (as opposed to those that are forecast). They are issued hourly, at 55 past, and cover tornadoes, lines of thunderstorms, embedded thunderstorms, areas of 40 percent or greater coverage of level four (very strong) thunderstorms or 3/4 inch or greater hail. These, or a center weather advisory which covers much the same thing, are usually on ATIS broadcasts; more importantly, controllers read them blind on the frequency or at least tell of the existence of a SIGMET for the area. They are often read rather rapidly, but the wise pilot listens to the points used to outline where the thunderstorms are located and, if any of the points are nearby, goes to the trouble to plot it on the chart. I wasn't en route yet, but while taxiing one day at Cincinnati I got a convective SIGMET that outlined an area of thunderstorms straddling the route of flight I had just gotten a clearance to fly. Before takeoff I gave them a new route, got a new clearance, and had a relatively uneventful flight.

Compared to the good old days, there has been progress in the air traffic control system dealing with thunderstorms. I remember the chaos that developed one day in the '60s as I listened at home on a VHF receiver while a line of vigorous storms moved through the New York area during the afternoon rush hour. Pilots wanted to deviate hither and yon, yet because of the saturation of traffic the controllers were having trouble sorting out where they could let who go. Some pilots were telling the controllers where they were going, which added to the disorganization of the melee. That it was all sorted out without bending metal was amazing. Since that time the FAA has instituted a system that will limit incoming traffic in bad weather or effectively shut down an area or route if violent weather is expected to have a significant impact. Instead of dispatching the normal supply of airplanes to a stormy Gotham, the controllers will cut the number—even to zero in extreme cases— so they don't have to deal with a significant amount of traffic and thunderstorms at the same time.

Impact on GA?

The impact on general aviation airplanes is minimal unless you happen to be going to a major metropolitan area that is affected. In fact, it can result in extra flexibility if you are flying to a nearby area simply because there will be less air carrier traffic and the usual number of controllers will be on duty. They will have more time for you, or, if you make a wise decision and land short of the storm area, you can probably get back up and on the way after the storm clears but before the inbound rush of traffic builds back up. It is honestly amazing how a wait of an hour or so on the ground will completely change the thunderstorm picture between here and where you want to go.

The FAA will occasionally reroute traffic because of thunderstorm saturation along a route; when this is done, we are procedurally separated from an area of bad stuff, but we may also be delayed as they sort things out.

Thee and Me

The big mean systems, though, are not the challenge most often faced by general aviation pilots. Where we have to work hardest is when there are storms out there and they are not impacting a major area or route to the extent that air carrier operations are curtailed. We do have to face the fact that in light airplanes we have to cull weather that the highly wing-loaded and much heavier jets can handle. Their big thunderstorm problems have been found just after takeoff or just before landing, where wind shear has precluded further flight. The big airplanes do get beat up en route but they tend to survive. A friend, an experienced pilot, told of a thunderstorm encounter as a passenger on an airliner where a flight attendant was bounced off the ceiling. Objects were thrown about the cabin and my friend was bruised by a flying handbag. In addition, his legs were bruised where they were banged against the bottom of the seat in front of him and there were also bruises from the safety belt. This says more about the integrity of Boeing airplanes and the crew's ability to fly in a bad situation than it does about the crew's ability to interpret radar.

In general aviation airplanes the biggest area of concern has always been enroute encounters with thunderstorms. The probabilities are just so much higher. For departure and arrival, we have a circle of maybe five miles we want to be thunderstorm-free. But draw the line from here to there and we see a large area in which we have to find a path that is free of thunderstorms. The higher flying jets might go over the storms; if they need to go around, a long deviation might take a third the time it takes us. Their radar is far better (because they use bigger antennas) than the best found in singles and light twins. And their airplanes simply do better flying in convective activity than ours. The advantage is all theirs en route.

Avoidance is the key, and as we study the information available after takeoff, all factors have to be balanced. One factor that looms large in many areas of the country is restricted airspace. When planning a deviation, beware the trap where you get caught between a restricted area and thunderstorms. It might turn out that the restricted area is as impenetrable as the thunderstorm. In such a case a pilot could and should exercise his emergency authority as pilot-in-command and tell the controller that to avoid the storm, he has to penetrate the restricted area. There may follow a lengthy period of explanation, but at least the pilot would be around to explain. A better plan would be a deviation in another direction. It might take longer, but perhaps not as long as it would take to explain the quality of a plan that caused the whole problem to develop in the first place.

What Isn't Seen

Even when the airplane is equipped with airborne weather radar and a Stormscope, and when the controller is helping with what he sees on his radar, what we see with our own eyes might count for more than anything else in getting the best possible ride. As cumulus build they reach a state where they can thoroughly beat up an airplane before precipitation starts to fall and can be seen on radar, or before lightning starts that would be detected by a Stormscope. We always want to deviate around these clouds, and it is best to inform the controller as far in advance as possible. If the deviation is to be only a few miles either side of course, chances are there will be absolutely no problem. But if it is to be a big deviation, the controller may have to coordinate to get approval. Altitude changes, too, should be requested as far in advance as possible. If we fly along for fifteen minutes with the thought that an altitude change might be needed, but without conveying this to the controller, it only makes everyone's job a little more difficult. If the airplane starts getting hammered, the flying is more difficult. If the altitude change requires coordination, the controller has to work harder. And there might be some reason, such as other traffic, why he won't be able to approve the change.

Low Altitude

There are times, too, when we have what we consider a very special requirement. This happened to me twice in just over a year, once at Dallas and once at Little Rock. In both cases a cold front had just passed through and was to the east, with scattered thunderstorm activity. You guessed it, I was headed east and would have to pass through the front. In the old-time tradition, I wanted to stay as low as possible because the thunderstorms had appeared to have high bases. The trouble with the air traffic control system was that there was no traffic control radar coverage in the area where I wanted to fly at 3,000 feet. Controllers are reluctant to let you use such airspace, but in both cases they understood what I wanted to do. I had the foresight on the trip out of Little Rock to request a block altitude of 3,000 through 5,000 feet before flying through the front, so we wouldn't have to worry about that, and we pretty well used up the block in the passage. There was no lightning on the Stormscope and no indication of a thunderstorm on the radar, but when transiting a frontal zone there's no way to get through without seeing a little action. On the trip out of Dallas, I forgot to request a block and used it up anyway, but got back to my altitude promptly. In neither case was the controller really able to give me any good advice about the weather, though on the Dallas trip the way the controller said it looked best appeared to have lightning activity on the Stormscope.

The most important thing to remember is that when dealing with active weather systems, the chances of getting a really smooth ride are not good. And turbulence not associated with thunderstorms can cause control as well as comfort problems in light airplanes.

Occlusion

Some of the worst turbulence away from active storms can be found as a front occludes, as a cold front overtakes a warm front. This usually happens as a low deepens, or strengthens and reaches its peak strength. As this happens, the movement of the low usually slows down and the circulation around the low intensifies. The wind shear involved in this process is considerable.

I was headed west one day with forecasts that didn't mention an occlusion, but the weather had taken a worse-than-forecast turn that day. The appearance of the clouds above was unfriendly. They almost looked like mammatus clouds—those dark-bottomed torn-up-looking clouds that always spell turbulence. As I flew along I kept requesting and getting lower altitudes, to stay below the unfriendly clouds. But I finally reached a point where I was as low as I could go. It started raining and became quite turbulent. This all happened near Dayton, Ohio, and the controller's situation was compounded by the fact that Dayton is one of those "hub" airports and a traffic push was occurring along with the occlusion. One airline pilot said, "Cells are popping up all over, right in front of us." It was indeed turbulent, but they weren't really thunderstorm cells. There was no lightning. And although the turbulence was bone-jarring, you could tell it wasn't convective because there wasn't a lot of up-and-down action. It was all related to strong wind shear.

I decided that it was time to land and requested an approach into Dayton. That was just what the harried controller needed, another airplane. He said I'd be about number 14 for the approach. I didn't relish flying for that much longer, so I requested a clearance back to Columbus, which I had passed earlier and where the weather was still pretty good. The interface between the weather and the IFR system at Dayton was simply not a good one.

What Gives?
On the ground at Columbus I checked weather, and indeed all the forecasts had been amended. And there were many pilot reports of severe turbulence around to the west of Dayton. But after a few hours' delay the occlusion had lost its punch, and I had a smooth flight on to Wichita, Kansas. Late arrival, sure, but when the weather is worse than forecast, and it is obvious that something dynamic is happening, those unscheduled stops are worth whatever time it takes.

Frosty Times
Another big weather interface with air traffic control comes as we try to deal with ice. There is a big difference here. To begin, it is quite difficult to know what the ice possibilities are by looking at forecasts. Whenever it is cloudy the forecasts call for icing in the clouds and in precipitation above the freezing level except during the dog days of summer. The forecasts are simply of very little help because they always include ice. Of a little more help are the freezing-level forecasts, but these often don't take factors such as temperature inversions into account. Finally, there is some apparent lack of understanding on the part of whoever devised the terminology for icing. "Icing in the clouds and in precipitation" is a misnomer except in the case of freezing rain. Otherwise, the ice is all in the clouds, caused by supercooled water droplets.

On days when ice is forecast, there are always a lot of general aviation airplanes out flying IFR. The majority of them probably do not have deicing equipment. And although there are from five to ten serious ice-related accidents every year, most of them occur after the airplane is out of the ice and while the pilot is maneuvering a well-iced airplane for landing. Occasionally there is what might be considered a classic ice accident, where the airplane gets so well coated it can no longer fly.

The air traffic controller's role in this is twofold. One, he is the man who can approve altitude or routing changes. Two, he can pass along information from other aircraft.

Ice is different from thunderstorms, too. If we wind up inside a storm, there is nothing the controller can do that will really assist the pilot. Neither an altitude nor a routing change will really help. All a controller might be able to do is tell you how many miles you have to survive in order to be out of the storm. In ice, though, he has the magic, the ability to clear you to change. And if ice is treated as it should be—any sign is a mandate for a new plan—then a pilot is not likely to wind up an ice statistic by continuing in icing conditions. It might cause the controller a little work and coordination if you call and request a change in altitude and a reversal of course to fly back to where there was no ice, but any controller would rather handle that for you than handle the emergency that might develop if a flight goes on into icing conditions.

True Confession

It was mentioned earlier that pilots might be reluctant to say anything about ice because of an apparently aggressive FAA enforcement program aimed at pilots who fly into ice without the proper equipment. This unfortunate situation complicates the relationship between the air traffic controller and our dealing with ice. How it is handled is an individual decision. What we can't do is allow it to slow down a request for a change. Saying "ice" may hasten the controller's work in getting us a new altitude or route, but perhaps if we are forceful enough in a request the message gets through without the magic word.

En Route Checking

Doing the proper en route weather work and overhauling the plan if necessary can address potential problems later on. One of the things that has always stood out to me is how wind can dictate a change in plans. Especially in the wintertime, the winds aloft forecasts seem to err, with the wind usually stronger at a lower altitude than forecast behind a strong cold front. Also, the wind velocity tends to peak wherever the air smoothes out on climb, and after that it increases gradually with altitude until you start getting up into the high teens where the jet cores start to come into play. The forecasts don't usually read this way. Turbulence aloft, above the low-level stuff, is always a good measure of a change in the wind simply because changing wind makes turbulence.

None of us like to fly in the bumps, so when westbound in the wintertime we usually climb into the higher velocity wind that is operating in the smooth air, usually above a cloud layer. I have always used a rule of thumb on this: If the area forecast says the tops are to be at 12,000 feet but they are at 8,000 feet, then the winds at 8,000 feet will more likely be near the velocity that was forecast for 12,000 feet. One other wind item: Wind velocity is usually strongest in a frontal zone and becomes lighter as you fly away from the front. The higher velocity in the frontal zone does not seem to be considered in the forecasts, so allow for a little extra there.

Why?

The reason all the wind business is important is that, westbound in the wintertime especially, we have the same endurance in hours and minutes but our range in miles is drastically reduced. Poor flying weather is often widespread, which can make for substantial questions toward the end of flights. What's the interface with the air traffic control system? Well, if you get to Indianapolis with it 200-1/2 though it was forecast better, with minimum fuel aboard because of a stronger headwind than forecast, and there is a string of airplanes ahead of you on the approach, you might be in a world of hurt. What better en route work would probably have done was resolve any potential conflict by causing a landing somewhere else, with healthier readings on the gas gauges.

One of the primary items in relation to weather and the air traffic control system is knowing, when the weather is low, where airports with ILS approaches and minimums are in relation to your position. Having an awareness of which of those airports might be impacted by a lot of traffic helps, too. We talked about Ohio a while ago. Dayton might have a lot of traffic, nearby Springfield not much. At Indianapolis, the big airport might be jammed while Mount Comfort, to the east, with an ILS, might involve no delay. Look at it this way: The airlines have to try to go to the big airports. In general aviation IFR flying, we have so much more flexibility and can choose alternate airfields that simplify the interface between the weather and air traffic control considerations.

How to Check?

What is the best way to check the weather when you are en route IFR? Flight Watch or an FSS immediately comes to mind. This works, but you do have to ask the controller to be off the frequency. Often he will tell you to report back on in five minutes. Then Flight Watch is busy with someone else or doesn't answer quickly because the specialist is otherwise occupied. So you watch the five minutes pass and then go away. At times the center controllers will help by getting weather for you when they are not busy, but they likely have weather only in their center area.

We mentioned listening to broadcasts while flying. If the ones ahead sound like death and destruction, perhaps you can hear one behind you. The uplink and display of weather will add a new element to checking weather as pilots install this equipment.

Analyzing the en route weather-checking process when the elements are misbehaving becomes a short-term process of nowcasting, as they say in the weather trade. What is going to happen now? What is going to happen in the next few minutes? Entirely different considerations are involved in what we learn about weather on the ground and checking weather in flight. Aloft, our only choice is to continue the flight to somewhere and if, as we fly, we always have somewhere good to go, then the weather requirement is satisfied. Forecasts are not of that much use in flight simply because they are not exact statements, and your flight has to be continued and completed while using only an appropriate portion of the fuel in the tanks. So if the destination is below minimums but forecast to improve, and the weather somewhere else is above minimums, the right decision might be to go to the good place.

A Sense

After flying IFR for a while, pilots develop a sense of how conditions are developing around the airplane as it plows through the clouds. The way this is done is by carefully putting together a picture of what you think it will be like before you take off and then carefully comparing actual conditions with the forecasts. Except on severe clear days, there will be differences. Then it becomes the pilot's job, while flying, to evaluate the differences and, if justified, to change the plan. Probably the greatest confusion comes to the IFR pilot when he tries to imagine that the weather is behaving as it was forecast when it really isn't. This is an area, too, where experience is worth a lot. By getting out there and getting wet the pilot learns the good IFR weather lessons. You can gain some of this experience by keeping up with the elements, whether flying or not, but, in contrast to VFR weather, you can't tell as much about IFR conditions from cloud and map watching.

One caveat about experience: If a pilot ever reaches the state of mind where he thinks that he has seen it all, so long.

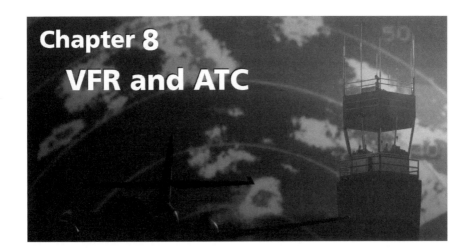

Chapter 8
VFR and ATC

Many people insist that dealing with air traffic control, even for VFR flights, has become so complicated that nobody wants to do it anymore. It has become more complicated, but as I was discussing this with my father, then 88 with 60 years of flying, he said "…but you can still take a compass and a clock and map and go most of the places that I used to go…" That is correct, and will likely remain true for a long time. But if a pilot is to fly in and around busy areas he has to learn to deal and live with air traffic control.

The FAA will likely always be in a mode where they will redesignate and rename regulated airspace, but it will always take much the same form that it takes now. There will be something resembling Class B or Class C airspace around busy airports, there will be places we can't fly without certain equipment, and the requirement to be in contact with ATC will be extended to more and more airports. One impromptu part of air traffic control, without controllers, will likely be expanded. The common traffic advisory frequency at uncontrolled airports could and should be improved, with additional frequencies for busy areas. There is so much congestion where there are a number of airports using these frequencies that the value is compromised.

Advantage: VFR
When dealing with ATC, especially around Class B airports, at times there is a distinct advantage to VFR operations. The controllers have more flexibility in dealing with VFR aircraft, both on vertical separation and on routing. The reason for this is that VFR separation standards are

less and VFR flights are not constrained by "letters of agreement," which cover IFR operations. We'll discuss these in the next chapter, but basically they provide for routings and preferred altitudes for flights moving between airspace controlled by various facilities.

It is important to have a plan before takeoff, but like so many things in flying, dealing with ATC on a VFR flight might require a continuous change in plan. And if you are working with controllers, they are very much a part of the plan and any changes that are made.

BWI/Philly

I was at my home base at Carroll County, Maryland, headed for Trenton, New Jersey. I had an IFR flight plan on file, but, as on so many routes in the northeast, this run is better flown VFR if you wish to fly the fewest possible miles between the two points. And although the flight was between Carroll County and Trenton, from an air traffic control standpoint it was between Baltimore and Philadelphia. They control the low-altitude airspace in the 109 miles between the two points.

I fly IFR almost all the time. For one, it is easier. For two, it accomplishes separation from air carrier aircraft, which to me is a desirable thing to do. If flying in a busy area like this, the air carrier separation can also be accomplished by getting VFR traffic advisories. That day I called Baltimore soon after takeoff, told them I had an IFR on file but would like VFR advisories direct to Trenton, and they agreed to provide this service. Their doing so did not guarantee that Philadelphia would also provide the service, so I had a plan that would have avoided the Philly Class B airspace as well as the feeder fixes I know they use. If you are not familiar with an area, it's hard to know all the feeder fixes, but there is some logic to them. Many terminals have a four-corner system in place— northwest, northeast, southeast, and southwest—where all arriving aircraft come in over one of the four feeder fixes. The altitudes vary, but jets are likely down to about 10,000 feet 30 or more miles away from the primary airport. If, that day, radar service was unavailable from Philly, I planned to go just north of the Pottstown VORTAC, 30 miles from Philly International, and stay at 7,500 feet. This would put me north and above Class B airspace, and hopefully out of the high volume of inbound high-performance jet traffic.

Nice Guy

Baltimore was doing good work, though, and gave me a handoff to Philly approach control. When you get such a handoff, the next controller should know who you are and where you are going, so the primary requirement in the first call is to give your number and altitude. "Philly Approach, Cessna 40 Romeo Charlie, level seven thousand five hundred." What actually happened that day was that he cleared me direct to the airport at Trenton and asked only that I call before starting a descent. Because I was above the Class B airspace I wasn't even required to be in contact with the controller, but I wanted to be. And for the descent, starting at a normal time meant that I would descend into the top of the Class B airspace and I had to have clearance to do that.

There was an interesting exchange between a controller and another VFR pilot as this flight unfolded. The pilot had elected to fly over the Modena VORTAC, closer to Philly than I would have gone. He had apparently not called Philly to begin. There were some scattered clouds in the vicinity of Modena, and something, perhaps the sighting of a lot of traffic, led the pilot to call Philly and mention something to the effect that he was having trouble seeing traffic because of clouds. The controller proceeded to politely ream the pilot out, making sure he understood that a lot of fast airplanes were out there and that if clouds were a problem it sure didn't make any sense to be there. The pilot responded that there were more clouds than he had expected and that he was having trouble staying a proper distance. The controller shot back that the pilot should say nothing more, he was only incriminating himself. I thought this VFR pilot handled the situation rather poorly and the controller very well. No action was likely taken against the pilot, but he was sent scurrying away from the busy area with tucked tail.

No Secret

It is no secret to air traffic controllers that some of the IFR routes prescribed by letters of agreement add a lot of flying miles to some trips. They are usually more than willing to oblige when you ask to do it VFR, as long as the weather is good and you aren't headed for an area where there is a high concentration of traffic such as a departure gate. When taking advantage of this service, remember that any vectors are for the purpose of keeping the flight as far as possible from harm's way. If you have a specific request, so state, but don't pout if it isn't met.

I have had some spectacular VFR flights in the New York area, courtesy of the controllers there, who are as helpful and friendly as any in the country. They might not sound quite that way to folks from out in the country, but the New York controllers are really helpful when they have time. We were flying back from Block Island, Rhode Island one day VFR because the only available IFR routing had been out over the Atlantic. I thought I would see if we could give our friends, not pilots, a special ride, so I asked the controller if we could go through the Class B airspace, over the top of LaGuardia, across Central Park, and then down the Hudson River at, if I remember correctly, 4,500 feet. "Approved as requested." There are traffic flows into New York that would have precluded such a flight, but that day it worked out and the controller seemed happy to approve it.

Busy vs. Boonies

There is a big difference in what we get out of VFR radar service in and out of a Class B area. The controllers know of every airplane within Class B airspace, or at least they know of every airplane that is in there legally. And they provide positive separation. But if you are out of Class B airspace and getting traffic advisories, you have no assurance of all, or even most, traffic being called. It's a service they provide as able, and it shouldn't be used as a crutch. The requirement to search out and avoid other traffic is just as strong as it is when you aren't talking to anyone.

In the Beginning

Even if you are launching on an IFR flight, it does not become such until the clearance is delivered. That might sound hopelessly basic, but I heard a pilot near Champaign, Illinois, fall victim to this. What he did was depart from a small uncontrolled airport VFR, to pick up his IFR clearance after takeoff. Trouble was, he entered the Champaign Class C airspace before calling departure control for a clearance. The result was a long period of swapping names and telephone numbers, and you can bet that this pilot flew on to his destination contemplating the 30-or-so-day certificate suspensions that he would surely get for this transgression. Just filing an IFR flight plan doesn't establish anything other than the fact that it was filed and that it will go into the system. The flight is VFR, and subject to all the VFR rules, until the clearance is delivered to the pilot.

Help

There are many stories on record where a VFR pilot or even a nonpilot was helped out of a predicament by air traffic controllers. But the call has to come before the help can be given, and the pilot has to realize the limitations to what people on the ground can do to help out. If you're up to your ears in unexpected or unfriendly clouds, or just lost, the best person to talk with is the controller sitting in front of a radar screen. Not only can he help you with position, he can also help keep you clear of other airplanes. The controller can't see clouds; staying in visual conditions is strictly up to the pilot. It is important to keep the controller advised of flight conditions, too. There have been cases where a controller trying to help a lost pilot thought the pilot was in visual conditions and could avoid terrain only to find out the pilot had been flying in clouds. What controllers know about terrain is the minimum safe altitude for the various segments of airspace. They don't have all the individual hills.

Various Levels

If a VFR pilot who is in clouds calls a radar facility, the first controller who talks with the pilot might be an experienced controller as well as a pilot, or he might be a brand-new controller and a nonpilot. After contact is established and the nature of the problem is clear, there is no question that any control facility would pull out all the stops to get the most experienced person in the room on the problem and to do everything possible to see the event to a successful conclusion. There may not always be an experienced instrument pilot around, and there are cases on record where another pilot in flight has taken over and helped the lost VFR pilot get on the ground. The point of all this is that a pilot in trouble had best be able to keep the airplane flying, wings level, and out of the trees while the air traffic control forces are being marshaled for the best possible help. Even given some time, the best they might be able to offer is vectors and information on safe altitudes. This, though, does offer the pilot an opportunity to devote all his attention to maintaining control of the aircraft. That is the important thing.

Approach Control

In the good old days the control tower was the VFR pilot's primary contact with air traffic control. Now, where there is radar and an approach control facility, the first VFR contact when inbound to the airport is to approach control. This should come after listening to the automatic terminal information service and getting word on the altimeter and wind and active runway, and on all the other stuff they cram into some ATIS broadcasts. They usually even give the frequency on which to call approach control.

What happens at some point after radar identification, in the opinion of some pilots, is that the controller will have you point the wingtip of the airplane, not the nose, at the airport.

What we have to understand, as pilots, is that the approach controller's job is to arrange traffic in an orderly flow to the active runway and then hand the aircraft off to the tower, all properly spaced for landing. It is true that the busier the airport, the more efficiently this appears to work. Consider this example at a relatively tranquil airport with only two airplanes inbound, though. Say a controller has a commuter turboprop on a 9-mile final for Runway 14 and you are flying a 120-knot airplane and are five miles northeast of the airport. Because the controller handles the commuter turboprops every day, he is quite familiar with the speeds used during the approach. On the other hand, some pilots slow a 120 knotter down to 70 when three miles out; others keep the speed up until closer or don't slow down as much. The controller has two choices — try to run the slow airplane in ahead of the faster one, at the risk of having to make the faster one go around, or turn the slower one to follow the faster airplane for a sure thing. Which would you do if you were the controller? Often they do what works best for them, not intending to discriminate.

Variables

Some pilots complain that controllers at some airports give priority to airline aircraft. There appears to be some truth to that, but it isn't or shouldn't be true because the aircraft are airliners. More likely it is because the airplanes are more predictable. Certainly flying my Cessna 210, I can fly approaches to within a mile or a mile and a half of the airport at any speed from 80 to 160 knots and still land. The controller may or may not know that the airplane has that capability, and he has no way of knowing how the individual pilot will use it. When they ask us to keep up the speed on final, they will often ask for a value that will be made good. Or, at times, they may just ask us to maintain a certain speed for as long as we can. It is still unpredictable how well we will manage the chore, though. And we have to always put ourselves in the other guy's shoes and cut him some slack on the way he handles traffic.

Tower

The basic function of the tower is to referee the runway space. There are, though, differences in towers and we had best understand them.

At an airport with an approach control facility, we discussed how approach control feeds airplanes to the tower on an orderly basis. If there is no approach control, the rules of the game are different. Pilots have to contact the tower before entering the airport traffic area, preferably when 10 or 15 miles out, but because there is no radar the tower isn't able to sequence aircraft. It can't do anything toward separation or traffic advisories either, other than to tell you someone else is out there somewhere. The clearance you get from the tower will be to enter the traffic pattern on a specified basis or to call a certain distance out on final. The only real control that is exercised is related to the runway. They'll ascertain who is first and clear that airplane to land. They also clear airplanes for takeoff, and when so doing will give advice on any other traffic they know about. The point is, when talking with a tower where there is no approach control service, our job is to stay alert and see and avoid other traffic.

There has been a plan in place to install radar at VFR towers but this has been slow in coming. The tower at my current home base, Hagerstown, Maryland, actually has radar in the tower cab but as of this writing it had not been formally commissioned. The result is that they can't officially use it but they have been telling of other traffic if it is close. When the use of it is formalized, I presume they will be able to use the radar to sequence traffic. Radar in VFR towers will help a lot because there have been a number of collisions at such airports.

No Guarantees

Don't take the presence of a tower or clearance to land as a guarantee of anything. One day a number of years ago, I was cleared to land on Runway 31 at Little Rock's Adams Field. As I was on final I saw an RF-84F on an overhead approach to Runway 22. My runway crossed that one at about midpoint. I asked the tower if I was still cleared to land. The answer was affirmative. I asked the tower about the jet, which was arcing around to final. The tower said there was no jet. It was apparent something was going on that the tower didn't know about, and at about the time I started a go-around the tower saw the jet, which had had a radio failure and was probably low on fuel, and started to sort things out. It is no sign of distrust to always question everything, it's just a double check. Another wise move is to always check final when you are cleared for takeoff, or before you taxi into position and hold.

The Language

An important part of dealing with air traffic control VFR or IFR lies in how we speak. Communication makes all the difference in the world. If you sound like a pro, they will treat you like one. If you sound frazzled and shout at the mike, they will treat you like a rank amateur.

So how do we sound like a pro? To begin, you need the proper equipment—meaning a noise-attenuating headset. To fly without one and an intercom is simply foolish. For a few hundred bucks you can make the airplane quieter and make communications with your passengers and the folks on the ground easier and better. I flew without them for years and regret it. Now, when I have to fly an airplane without a headset and intercom, I wonder why in the world I ever did without.

The reason a headset helps so much is that it reduces the ambient noise so you do not feel like you have to shout to be heard. If the radio equipment provides side tone, it gives the added benefit of hearing yourself talk, and all of us like to sound good to ourselves. You really will tone up your transmissions and measure your words when you can hear yourself. The language of aviation is nice, and as you practice your delivery you come to enjoy it.

One word on headsets: Some pilots reject them because they say they are not comfortable. It is true that what is a comfortable headset to one pilot feels like a set of ice tongs to another. There is, though, a headset out there that will be comfortable for you. Try them all to find the one that fits the shape of your head. Being square-headed, it took me a while but I finally found one.

The Message

There is plenty of material around on what words you should use to convey given messages, but this isn't really something you need to memorize. The main thing is to remember to be concise, to be clear in meaning, and to avoid jargon such as CB talk. I heard a pilot say to a controller that he would "cancel IFR and go VFR if it will help you out." That isn't a clear statement; it is actually a question. In this case the controller took charge of the situation and simply acknowledged the pilot's cancellation of his IFR clearance and went about his business.

Don't wind up like a pilot back in the good old days, who was apparently confused about terminology. It was a really grungy day when his voice came through on a tower frequency, a little squeaky, indicating some trepidation, and asked for landing instructions. Because of the weather the tower shot back, "Are you VFR or IFR?" The pilot replied hesitantly, "Ahh, gee, no, I'm on the GI Bill out of Teterboro."

Some controllers stick very closely to the prescribed terminology, others are not so precise. And that is where we should get our clue. If the controller appears to be carrying on an informal conversation and the frequency is not busy, we can be easy. If the controller is quite busy, and is totally crisp and businesslike, best follow suit.

One of the biggest complaints about VFR communications is the long-winded speech. A pilot calls, the controller answers, and then the pilot comes on with a quite detailed description of what he is doing, where he is going, and what he wants. If the controller happens to have a complex IFR condition developing, this effectively shuts him off from his primary job for the length of the speech. The best VFR procedure is to listen on the frequency for a while, find a quiet spot, and then say something like "Cleveland Center, Cessna 40 Romeo Charlie, you have time for VFR advisories?" That gives him two good choices: He can either say negative, or, if he has time, he can solicit the information he needs to provide the service.

If the communication relates to a problem, be up-front about it. For example, if you have some sort of an emergency and feel that the controller can offer assistance, don't be bashful. Tell him you have an emergency. It was not a VFR flight, but the fuel exhaustion accident involving the Avianca 707 will be used for years as an example of how pilots apparently understated their case. In this tragedy, the fact that English was not the native language of the crew has to have had some effect, but still, had they screamed and hollered early on and declared a fuel emergency and told the controllers exactly what they were capable of doing, some aid might have been available. We will explore this more fully in the next chapter.

The Beehive

Although it is honestly easier to use busy airports IFR, there are times when we need to use them VFR. If it is an airport with slot restrictions, such as Washington National, there might not be an IFR reservation available, leaving VFR as the only alternative. The difference between the two shows why VFR is more difficult and outlines what we have to do to fit in smoothly.

When arriving IFR, they know you are coming, you are handed off to approach control from the center, and no introductions are necessary.

When you want to go in VFR, you have to find time on the appropriate frequency to get a word in and tell them who you are, where you are, and what you want. In the case of an airport in Class B or C airspace, they can simply tell you to remain clear. Or they might clear you into the airspace and work you into the traffic flow to the airport. Controllers at busy airports are masters at working a lot of traffic and they are quick to judge how well a pilot will fit in. Needless to say, the ones who sound as if they will fit in get the best service. Communicating concisely and following instructions to the letter suggest that a pilot will be relatively easy to fit into the big picture. Hesitation and misunderstanding suggest the opposite. And, really, we should not try to fly to big and busy terminal airports unless confident in our ability to do so without causing a basic disruption of traffic. Some very experienced pilots are not comfortable in major terminal areas. Nothing to be ashamed of there, it is just a suggestion that until they take the steps necessary to become comfortable with busy places, best remain clear.

Familiarity

I have flown in the Washington-to-Boston area for a long time, feel quite at home with the big airports in the area, and have used all of them at one time or another VFR and IFR. In advance, I know basically how they are going to handle a VFR flight, where they want you to go, and how they want you to behave. I have talked with the controllers enough that I think I understand their problems. I can accommodate their request to go fast or go slow or to land on a runway with a crosswind, or if I don't like it I can just go away. I go to these airports with a strong desire to fit in with the least possible work on the part of the controller. In other areas, I don't feel as comfortable and, in fact, studiously avoid VFR into unfamiliar major airports because IFR works so much better. If it becomes absolutely necessary to fly VFR to one, I study the area in advance and try to imagine or anticipate how they will handle the arrival. For example, I haven't flown into Dallas/Fort Worth VFR enough times to have any reasonable idea of how they move traffic and am even perplexed about why I am sent where I am sent when going there IFR.

If you're going to a busy airport, call in before you get too close to the edge of the regulated airspace. One mistake that none of us should make is to blunder into regulated airspace without establishing communications and getting the proper clearances.

Go See

Finally, one thing all pilots should do is visit air traffic control facilities. We should all "know" some air traffic controllers. To know them makes the relationship a better one. It's a good idea to have one come and talk if you have a local pilot's group, or the FBO might invite a controller to come and talk to a ground school class. Knowing controllers and visiting facilities helps in the understanding of airspace.

There is very little uncontrolled airspace left in the United States or, for that matter, in most of the developed nations of the world. Whether the rules require that we be under control or not is a moot point. The fact is that somewhere a person is overseeing the movement of IFR traffic in that airspace. That we are free to use it VFR is wonderful, but we do need to be aware of the problems of the controllers and limitations on what we can do and where we can go VFR. If you live near a major area, knowing the traffic flows into and out of the hub airports helps minimize conflicts with fast traffic.

Chapter 9
IFR and ATC

IFR flying gives us our greatest interface with air traffic control. It is important to understand the limitations on the other guy here, too, lest we continuously get crossways with the system. Often we hear other pilots grumbling on the frequency, as if the controllers were just arbitrarily sending the aircraft out of the way. In fact, the flows of traffic are carefully planned, and in major metropolitan areas the controllers can hardly cater to the whims of each individual pilot. There is a procedure for everything, and only by following those procedures can they handle the traffic.

Some things, however, are not logical. For example, transitions are published on instrument approach charts. I flew from Mercer County Airport in New Jersey for years, flying countless ILS approaches to Runway 6 at that airport and never ever flying the published transition for that procedure. There's nothing wrong with that, the arrivals were always vectors to the ILS, and even for perfect strangers there was no problem. This does make you wonder why they publish the transition. It also illustrates that even though you may plan your arrival to an unfamiliar airport carefully, it's best to be ready for changes in plan. Simply put, we do as they say. That is the basis of the air traffic control system. Otherwise there would be chaos.

IFR Flight Plan
When we file an IFR flight plan, whether over the phone or with a personal computer using the direct user access terminal system or private service, the process is not very automated. The computer will check

some things about the routing, but it does not possess overwhelming logic. If all the fixes used are valid, and if the airways specified cross all the fixes, the computer will accept the flight plan. The fact that you file direct from over Martinsburg, West Virginia, to a station in South Carolina and then direct to Chicago won't likely be questioned even though that is nowhere near a direct route. Likewise, anytime the computer gets indigestion over an airway you have selected, even though it is depicted on the chart, by filing direct between the two points in question, you satisfy the computer. At times their computer will add a point to your flight plan. Flying out of Carroll County in Maryland, using V3 to go to Martinsburg, it always adds Ruane intersection, between Westminster and Martinsburg VORTACs, as a reference point. This is, according to the computer, for ZDC's purposes, ZDC being the code for the Washington Air Route Traffic Control Center. Yet when I call on the phone to get my clearance from Baltimore approach control, I have heard them discussing why this yo-yo at Carroll County filed for the Ruane intersection. Your computer, folks, not mine.

Geo Lesson

The computers often appear not to understand geography, too. The flight service stations use basically the same program that runs DUAT to give us weather when we call them. The weather information program simply did not know where the base airport was located. If you asked for a briefing to Carroll County, W54, on the phone or through DUAT, the closest weather it would give is Martinsburg, much farther from W54 than Baltimore, which also happens to control IFR traffic into and out of W54. Further, Martinsburg is on the other side of the first ridge of the mountains, which often means the weather there is substantially different than at W54. Slaves to the computer that we have become, most FSS folks will question why I asked for a weather briefing to Baltimore when really I was going to W54. I tell them why and they respond that this couldn't be the case. Then I hear the tap of keys and they come back and say, "You know what, you are right." The point of all that is to illustrate that we can't expect the computer to think for us in regard to routing or to weather.

Routes

One thing computers will do is adjust routes, and at times they do this in what may seem a mysterious way. For example, when I am flying to

Asheville, North Carolina, from my base in Maryland I usually file over Martinsburg and then direct to Barretts Mountain in North Carolina, V222 Sugarloaf, direct to Asheville. That flies in the face of one accepted rule—always file over a high-altitude VORTAC in each center's airspace. Every center has all these VORTAC locations in their database. They have locations of other than high-altitude VORTACs only for their own airspace. Neither Barretts nor Sugarloaf is a high-altitude facility. So my direct Barretts routing is seldom issued as filed even though it is accepted by the computer. They usually read a full airways route clearance. But after contact with Washington Center, they often approve direct to Barretts Mountain. Apparently the controllers know where it is if the computer doesn't. On the return trip, the route direct from Barretts Mountain to Martinsburg is always approved. Why? Because Atlanta Center knows where I am to begin with, and they know where Martinsburg is because, while in Washington Center's airspace, it is a high-altitude facility and is thus in Atlanta's computer. Often if something doesn't appear logical a little study will shed light.

Lat-Long

One way around this is to file to the latitude and longitude of the facility in question. This can work because the computer understands lat-long numbers. Although some pilots are comfortable with lat-long, I never did like to use it. My first LORAN required lat-long inputs, and when I got my first Northstar and no longer had to fool with lat-long, I felt liberated. Like the first day I realized I no longer had to remember Morse code to be a real pilot.

However it is done, if you want a direct route, file for it. The process is easier on everyone if the flight plan is filed as the pilot wishes to fly the flight. Then if ATC has a different idea, it is their prerogative to change the routing.

One other item on filing: If you are headed to a major metropolitan area, and if no Standard Terminal Arrival Routes are published for the area, go ahead and file direct to the airport and see if it works. It isn't likely to work if the airport is a satellite of a major airport. For example, the feeder fix for the satellite airports in the Philadelphia area from the west is Harrisburg. They'll clear you direct to there from anywhere, but after Harrisburg you have to follow the prescribed route. Going to the general aviation airports in the Kansas City area from the east, the magic spot is Napoleon VORTAC.

Air Possession

Earlier I mentioned letters of agreement between the various traffic control facilities. These prescribe who is responsible for what, and outline the preferred routes. Exceptions can be made, but they require extra work on the part of the controllers. It is really in our best interest for their workload to be as low as possible, too, because when exceptions are made, they occasionally lead to mistakes.

It helps to think of the facilities as kingdoms (centers) and fiefdoms (terminal radar control facilities). The centers are responsible for large-scale traffic flows and control all the airspace above certain altitudes all over the country. The terminal facilities control airspace near airports, below certain altitudes. When we fly low-altitude IFR, we fly a lot in airspace controlled by terminal facilities as well as along a network of low-altitude routes where you never talk to a center. As a rule of thumb, terminal facilities control the air 10,000 feet and below within 30 miles of the primary airport, though there are many exceptions.

Operationally, the division of airspace can have an effect on how our flights are handled. If, for example, you are overflying a terminal in airspace controlled by a center and you wish to descend to a lower altitude for better winds, it may take a while to get the approval for a descent. If the center controller does not "own" the altitude you want, he would have to coordinate your descent with the terminal facility.

Altitudes

In recent times, with various airspace overhaul plans, we have seen an increasing use of lower altitudes for aircraft flying to satellite airports in major areas. This is a way of segregating traffic going to satellites from traffic going to the primary airport. It is quite effective at traffic separation, but it puts the satellite traffic down in the low-level turbulence for a while. If you happen to be flying a turbine aircraft, the fuel flow is quite high as well. Perhaps the best thing to do is just plan on whatever extra fuel is necessary. I have tried to beat the system when it is VFR and go in using a normal descent profile. Every time I do it, though, I find the level of heavy high-performance traffic to be a touch on the high side. The reason they want us to fly low is indeed related to the traffic volume. In the area to the west of my home base, over near Frederick, Maryland, the inbounds to Dulles from the north and from overseas are quite numerous and a VFR descent betwixt and between them is not comfortable.

Another altitude that comes into play if you fly a turbocharged or turbine-powered airplane is Flight Level 230, 23,000 feet with the altimeter set on 29.92. Above this altitude the high-altitude sectors take control of the airspace in most areas. Out in the country you shouldn't have much trouble getting altitudes in the mid to high 20s, but in busy areas FL230 might be the highest altitude you will have approved unless flying a jet that fits into the high-altitude traffic flows. Even then, some routes have been limited to altitudes below FL230, regardless of the performance of the airplane. This means more fuel and it's just a byproduct of increasing traffic, enhanced by the rapid proliferation of regional jets. Usually when you hear horror stories about routing and handling from pilots flying relatively slow turboprops, they were trying to operate above FL230. Still another altitude is FL290, above which 2,000-foot vertical separation is provided. Despite the constant references to 30,000 feet in the movies, that is not an active flight level. FL290 is eastbound, FL310 westbound, FL330 eastbound and on up, switching each 2,000 feet. As this is written, there is a plan to reduce that high-altitude vertical separation to 1,000 feet, which will help a lot en route but it'll take more runways to alleviate congestion at the airport.

Weather

When we are flying in the air traffic control system as well as in a well-organized inclement weather system, we have to recognize the strong role that one plays on the other. Put yourself in the shoes of the Avianca crew who used all the fuel on a trip from Medellín, Colombia, to JFK airport in New York in early 1990.

The weather was affected by a complex frontal system related to a strong low over northern Lake Huron. The New York area was in the warm sector of the weather system, with low ceilings, rain, and gusty southerly winds. The JFK weather just before the crew flew an ILS approach was indefinite ceiling, 200 obscured, visibility a quarter of a mile in light drizzle and fog. The wind was from 190 degrees at 20 knots gusting to 28. The reported runway visual range was 1,800 feet variable to 2,200 feet. The weather conditions were having a strong impact on traffic at JFK. Some aircraft were held en route, others were held on the ground at the departure point, and still others were diverted to alternates. As much as we might like to think that the system is designed to work well in fair or foul weather, when the weather is bad, delays develop—especially at the busiest traffic times. According to an Associated Press report, on that day the traffic control supervisor at the New York TRACON had recommended that just 22 airplanes an hour be accepted at Kennedy. The central flow control facility in Washington was consulted, and the acceptance rate was set at 33 airplanes an hour—a number they apparently could not handle without rather extensive delays. It was also reported that on the tapes that evening an American Airlines flight transmitted that they had only enough fuel for the approach and landing.

The measure of trouble that the air traffic control system was having with weather was obvious from the progress of the Avianca flight. It had been held over Norfolk for almost 20 minutes, near Atlantic City for almost 30 minutes, and about 35 miles south of JFK for another 30 minutes as the controllers did their best to work it toward the destination.

While the Avianca 707 was holding 35 miles south of JFK, the controller called with a new time to expect further clearance, the second extension of the EFC time at this holding point. This one was about 20 minutes and a crewmember responded that they needed priority. In response to a question about how long they could hold and the alternate, the crewmember said that they could hold for five more minutes and that Boston was the alternate "...but it's, ah, full of traffic...."

At this point a New York Center controller called a New York TRACON controller and told him the flight could do only five more minutes in the hold. "Do you think you could take him or should I offer him his alternate?"

The TRACON controller responded, "Slow him to 180 knots and I will take him."

While the handoff to the TRACON was being made, in expanding on the question about the alternate, the flight crew said to the center controller, "It was Boston but we can't do it now, ah, we will run out of fuel now."

After the handoff to the TRACON, the flight received routine vectors and descent clearances plus one 360-degree turn for spacing. The crew was given a wind shear advisory and handed off to the final controller when the airplane was down to 5,000 feet.

At about this time the crewmembers had a discussion that related to the correct procedure for a go-around with low fuel. There was mention on the flight deck that they were being accommodated ahead of other traffic, and that the controller knew of the bad condition of the flight. The controller asked the flight for 10 knots more speed and had to ask again a few moments later because of a following TWA flight that was unable to slow down any more. The Avianca flight's ILS approach appeared to be normal except for some excursions below the glide slope; at one point the ground proximity warning system gave 17 seconds worth of "pull-up" commands. A crewmember remarked that they were in the wind shear at that time, about 500 feet above the runway. Then the captain asked where the runway was and the first officer replied that he didn't see it, so the captain started giving the commands for a missed approach. After the go-around started, the captain told the first officer to tell them the flight was in an emergency. This was transmitted as "...we'll try once again, we're running out of fuel." The captain asked again if the tower had been told they had an emergency; the first officer responded that they had been told. The approach controller was informed that they were running out of fuel and this statement was repeated a few moments later. The flight was 15 miles from the outer marker and had been cleared for another approach when it ran out of fuel.

The National Transportation Safety Board made much of the fact that this crew never declared an emergency. The controllers who were communicating with the crew said that the remarks about running out of fuel were matter-of-fact and did not convey urgency or an emergency.

Lessons

There are a lot of lessons in this tragedy for general aviation pilots. A foremost one is related to flexibility, which we have more of than do airline crews—especially crews of relatively small foreign airlines. We can go to almost any airport and get fuel with American Express or MasterCard. The crew of a Colombian 707 would be more motivated to go to an airport that has ground-handling support or an agreement for ground-handling support. From the first hold at Norfolk, it was apparent the flight would have been delayed. Flying your own airplane would have made it a simple matter to land at Norfolk and tank up on fuel. It would have been inconvenient, and it would have likely delayed the arrival at JFK because it is harder to get back into the system than it is to stay in the system when headed toward a busy airport on a stormy night. But one of the keys to managing risks that might be generated because of bad weather and air traffic control problems is the willingness to get the airplane onto any adequate runway if the margins and reserves appear in peril.

The fact that the crew discussed the procedure for a low-fuel go-around, and that they transmitted they could not go to their alternate, was a clear indication they were very aware of a low-fuel state. But "emergency" or "Mayday" was never transmitted. It was only mentioned by the captain on the flight deck. If they had declared an emergency before the first approach, they might have gotten to the runway a little quicker, but there is no suggestion they would have completed the approach any more successfully. Only about ten minutes elapsed from the time of the missed approach to fuel exhaustion. Had the word "emergency" or "Mayday" been spoken, they might have been able to get the airplane around for another try, but that is quite a short time to climb to a safe altitude, come around, intercept the ILS, and fly it to the runway. The lesson here, again, is to be very clear and specific when you have a problem. An emergency is a distress or urgency condition. This crew was clearly in that state, but they never came out and said so to the controllers.

It is never good to be out there flying an approach with no options — which was what this crew faced. If ever you back yourself into a corner like this, the only salvation is in finding that runway at the end of the ILS. In light airplanes, at slower speeds, the ILS is easier to fly precisely and it will actually lead you to the runway. Back in the good old days when flight directors first came out, the really sharp demonstration pilots would land under the hood. That is no suggestion to go out and bust minimums, which would be a last-ditch and rather desperate move. Faced, though, with fuel exhaustion over a major metropolitan area, this crew apparently had only this one other option left when they started the approach.

Control Ability

When flying in the system, we have to always have a clear picture of the capabilities of the people and the equipment on the ground. Perhaps one IFR item that is often overlooked relates to other traffic. The controllers can see almost all transponder-equipped VFR aircraft that are flying out of Class A airspace, which is below 18,000 feet, but they can't always see aircraft that are not equipped with a transponder, which is legal below 10,000 feet outside regulated airspace. The result is that other traffic may or may not be called. The controllers' primary duty is sorting out and separating the IFR flights; the work they do for us calling traffic is something done when time allows. It is simply our job to see and avoid other traffic when flying in visual meteorological conditions.

Weather Avoidance

Pilots frequently ask controllers about weather because controllers have a limited ability to "see" precipitation on their scopes. This ability is certain to be enhanced sometime in the future when controllers should have quite accurate overlays of weather on their scopes. This doesn't mean they will be able to make interpretations of weather for us, though.

What we see and feel, and the airborne equipment, will still count for much. The real benefit will be in their planning of the big picture and the help they can give us with the big picture.

A good example of the benefit of this will be in areas where there are large quantities of restricted airspace. Florida is a prime example. Usually, for some reason, the thunderstorms form in the airspace that is not restricted, and when we are faced with the zig or zag decision the options are usually to fly out to sea a ways or to try to squeeze between the restricted area and the storm. Even with airborne weather radar, we have to make these decisions 20 to 40 miles at a time. A controller with a good overlay can describe how it looks, and the best path, on a much grander scale.

Active Day

Flying to Oshkosh in 1989, with Captain John Cook in the left seat of my 210, we got a fine exposure to the interface between thunderstorms and air traffic control. Cook, who had just retired from British Airways and who took their beautiful Concorde to Oshkosh in 1985 and 1988, was getting a lesson in slow-motion weather avoidance, compared with what he had done with Concorde. I think he also got a lesson in how flexible our air traffic control system is, as well as how much airspace we have to work with in the United States. It was fun flying with him in my airplane after flying with him in both Concorde and the Concorde simulator. John, the quintessential English airline captain, flies with a deft touch and a deliberate manner, and had to really put the brakes on the thinking process. In Concorde you pick a good spot and dash through. In a 210 you look at the spot for a long time before you reach it.

There were actually three lines of thunderstorms along the way, plus activity in the Oshkosh area. On the third line of storms we reached, the controller's radar wasn't giving us useful information. He suggested a heading, we turned to that heading and examined the sky visually, with radar, and with the Stormscope, and opted for a somewhat longer detour. There was sure no debate in the cockpit of the 210 about the advisability of flying the extra miles.

Nor was there debate about a fuel stop at Muskegon, Michigan. They were reporting thunderstorms at Oshkosh and Green Bay on the other side of the lake, and although we had plenty of fuel to get there, we didn't have a lot of fuel to use holding. Holding was being done over there, too, because of the high volume of traffic flying in for the show and because of the thunderstorms. John was quite impressed that, en route to the largest air show in the world, we were able to deviate around storms as requested every time, and to make an unplanned stop, file a new flight plan, and get a clearance without delay. We were headed for Green Bay instead of Oshkosh, mainly because to me it is easier to go there and rent a car and drive to Oshkosh than it is to fly into the beehive.

Restricted Air

A nice feature of IFR flying is that the controllers keep you out of trouble on restricted airspace by not clearing you through it. The option might exist, too, for corner-cutting through restricted airspace if it is not in use. The controller would know. A famous restricted area is on a frequently flown route—V268 from Westminster VORTAC to Hagerstown VOR goes right over Camp David, which is restricted below 5,000 feet but is available for use at 5,000 and above whenever the President is not in residence. One thing you can't do IFR is get a clearance through an active Military Operations Area. Where it is legal, it may be very scary to fly through an MOA VFR, and they won't clear you through one that is in use when you are IFR.

GPS sets have a warning feature that covers regulated airspace. The user can select which warnings the set gives, though, and if a pilot suppresses the warning and then violates regulated airspace, that pilot is in deep trouble.

Contest

I have written a number of times about the instrument flying contest at the National Intercollegiate Flying Association's annual safety conference. The pilots are sharp, but occasionally you fly with one who has trouble talking with the controllers. We even had one who had never talked to a radar controller even though he had his instrument rating. Vectors came as a total mystery to him, and although he did a pretty fair job of flying the airplane on instruments, he had a terrible time fitting into the air traffic control system.

On the other hand, I flew with the winner of the contest in 1990, Todd Ericson of the Air Force Academy, who was quite at home talking to controllers and flying through clouds in the academy's 150-horse-power Cessna 152, complete with an Air Force One paint job. To work well in the system, you have to have confidence, and Ericson had that in just the right measure.

How to Speak?

There is a language to use in talking with controllers. The best way to learn the language is to study the books as well as listen to other pilots on the frequency. A VHF receiver at home can do wonders for your ability to communicate concisely and clearly. One other thing does wonders—smile when you talk to them. If the frequency is not busy, sign off with a "thank you" or a cheerful "g'day." Some pilots sound as if they are grumpy about being in the airplane and in the system. Flying is fun, though, IFR is fun, and by being cheerful about it we can honestly have a better relationship with controllers.

We need to work at the cheerful nature when something comes up that we don't like. Using an air traffic control frequency to complain is something many pilots are guilty of, and this is a very poor practice. If you run up on handling that you think is bad, or get a route that you don't understand, don't mention it to the controller. He is just doing his job. The phone numbers of all the centers and TRACONs are listed, among other places, in AOPA's Aviation USA. If there is a question about something, call after you land and they will be glad to give you an explanation. Differences of opinion about routing or handling are best handled this way. Just make a note of the time, the frequency, and your position and give this to them when you call.

Silent Treatment

When we are en route and get a frequency change, the new controller may not answer right away. There is always the possibility that the previous controller gave you the wrong number, or that you got the number wrong. The enroute charts show sector frequencies, so you can check the number there, or you can go back to the previous controller. But it is best to wait a few moments and then call again. The controller might have been taking care of something else when you first called and will acknowledge your presence as soon as he can. There is nothing time-critical about making that contact as long as you continue to fly your clearance and maintain the assigned altitude.

The good frequency news is that many GPS sets will tell you what the frequencies are for the area in which you are flying. The Garmin GNS 530, an extremely talented navigator, will even give you the bearing and distance to the remote communications outlets. I use this when the center transmissions don't sound too strong. When that's the case, I usually learn that the frequency I am using is farther away than normal.

Nice Time

The en route portion of an IFR flight is nice. As your airplane drones along, communications with the controllers is a good way to spend time. Adding some weather for spice makes it even nicer by giving you something else to think about. Just remember the old business about all responsibility being vested in the pilot-in-command. There are always two systems out there—the air traffic control system and the weather system—and both the responsibility and the challenge come from fitting the airplane into both of them as smoothly as possible.

Chapter 10
The VFR Arrival

So what could be a big deal about a VFR arrival? Simply put, it can be one of the most demanding maneuvers in flying to do smoothly and correctly. It is a phase of flight where planning can be everything.

One of the best examples I have seen of this came in Montana in 1971. I was flying with Jeff Morrison of Morrison's Flying Service in Helena. We were in my Cherokee Six and Morrison was giving me some good instruction on using mountain strips. The smallest and most confined one we used was Meadow Creek, buried in the mountains and located by a stream. It might even have been classified as a river. It was quickly apparent that this would have to be a well-planned approach. The final approach would have to have a bend in it, to follow the river valley to the runway. There would be no last-minute go-around because the other end of the strip was blocked by a mountain. Morrison was telling me what to do, what speed to fly, and when to extend the flaps. I was a voice-activated autopilot. A little to my surprise, we didn't finally configure the airplane for landing and slow to the reference speed until the landing was assured. Morrison's premise was that by flying a little faster and a little cleaner until we knew we had it made, we would have better options on any abort of the landing attempt. It made a lot of sense to me, and we were soon walking around on the idyllic strip, enjoying the view. An IFR arrival will certainly get you on the ground in grungy weather, but this VFR arrival offered a far better view and environment after landing.

There was no interface with air traffic control on that approach and the weather was clear. But there was a weather consideration, wind being weather. Earlier, flying in and out of another strip, we had encountered some turbulence that was a result of wind flowing through the mountains. On the approach to Meadow Creek we had to visualize how the wind flow might affect our approach slope. It wasn't a big factor, but there was some up-and-down action as we worked the Cherokee Six into the strip.

Go Now To...

The moral to that story is that planning makes any VFR approach and landing better. If we blow into the traffic pattern of a strange airport without making some plans, chances are the approach and landing will be anything but smooth and well done, especially if it is windy or weather is a factor.

I like to start planning for an arrival when I am about 60 miles out if it is a long cross-country flight and never more than 30 miles out. Some pilots like to consider the arrival as beginning when leaving cruising altitude. The main thing is to pick a time that is the beginning of the arrival, a time when you want to have studied the airport diagram to be familiar with any obstructions in the area, to have listened to the automatic terminal information service broadcast, to have selected the fullest fuel tank for landing, to have briefed any passengers for the landing, and to have generally tuned yourself up for a well-planned arrival.

On a VFR arrival the first call to approach control should be made well before reaching the regulated airspace. Even at a relatively bucolic radar facility, they control the airspace out to about 30 miles away from the airport and usually have radar coverage farther out than that.

One thing is certain when you are flying toward any airport: The density of traffic will increase the closer you get to the airport. This is just like en route, too, in your being responsible for seeing and avoiding other aircraft. The radar service in a terminal area is more complete than en route, and in the case of Class B or C airspace there should be no unknown airplanes flying around out there. But it still pays to look as a double check on the radar service.

Wind/Weather

The wind is important in planning the arrival, approach, and landing. Going into Tulsa, Oklahoma, one windy spring day with our son flying my 210, I noted a tailwind of over 70 knots at about 1,500 feet above the ground while we were on the downwind leg. The surface wind was reported as 30, with gusts to 48, which is a bit on the high side, but there was no way for us to stay up until the wind abated.

There would definitely be wind shear on this approach as well as more than a little turbulence. Fortunately, the strong wind was blowing down the runway. I told Richard what I thought the best plan: Approach using only 10-degree flaps and fly the approach at 105 knots. The reduction in flaps was to get a higher touchdown speed and to have less drag should we encounter a significant wind shear. The 105 knots was 15 over the normal for a 10-degrees flap landing, which I thought adequate to handle the gustiness. The airspeed always jumps around in conditions like this; it is always my goal to make the lowest excursion of the airspeed to be the normal approach speed or, with the flaps extended only 10 degrees, 90.

The tower was giving us wind checks as we flew final. I was a happy man, Richard was doing a good job of flying, and it seemed that despite the strong wind the arrival would be somewhat normal.

Then, in my peripheral vision, I sensed that the world was rising around us. A quick glance at the airspeed showed it to be on about 85. Richard had already started forward with the power and was flying a good pitch attitude, slightly nose up. Then we hit a significant bit of turbulence that made the stall horn beep momentarily as the airplane flew through the shear. The airspeed started building so the drill was to start getting off the power in order that we wouldn't reach the runway with excess airspeed.

Had we not thought through what might happen on that approach and made a plan that maximized the capabilities of the airplane, it would not have gone as well. Although light airplanes are generally better in wind shear than heavies, we do have to be wary. In light airplanes we have less drag and better acceleration capability in the approach configuration, but it's still good to take advantage of everything.

Weather/Weather

Checking the arrival weather on the ATIS as well as how it looks should give some clue as to what might be expected. But on days when there are passing showers or thundershowers, a VFR arrival has to be flown with a continuing willingness to punt if necessary.

A number of years ago my father, Leighton Collins, and I were headed for Washington National one day in his Twin Comanche. We were flying in the warm sector of a weather system with, as I remember, a ragged ceiling at about 3,000 feet, pretty good visibility, and some bumps. There was a strong southerly flow. This was back when the TCAs (now Class B airspace) had just sprouted and the reservation system for IFR operations into National was in effect. Lacking an arrival reservation, or slot, we were of necessity VFR.

Because I was a more frequent visitor to National than my father, he flew and I suggested which way to go. Flying in from the northeast, with a south operation at National, it always seemed best to me to fly down the Anacostia River which extends to the northeast from National, calling them at the appropriate place.

The airspace at National is confined by Andrews Air Force Base as well as by the prohibited areas that keep airplanes from flying directly overhead politicians. The Anacostia River is a fine navaid to use on an arrival from the northeast (as is the Potomac from the northwest or south). Follow it and you get navigational service that keeps the airplane clear of regulated airspace and also gets it to the airport.

National was reporting VFR on the ATIS and we got our clearance into the area with no problem: To follow the Anacostia for a landing on Runway 21, which is not far from straight-in off the Anacostia.

The closer we got, the more obvious it was that showers were moving through the area. I showed my father a landmark that would line us up with the runway and then, through a shower, saw the airport and the runway. It was his turn to be a voice-activated autopilot until he actually saw the airport and took us in for a flawless landing on Runway 21.

Not a Good Idea

I am not sure I would do that VFR today. The options we had that day were limited, and I don't know what the controllers would have done had the weather at the airport suddenly gone IFR. It seemed to us that the visibility went below three miles, but the excellent controllers at National kept handling VFR traffic. Certainly mixing an increasing number of showers with a substantial number of VFR airplanes, in airspace that is filled with restrictions, doesn't leave the pilots or the controllers with a lot to do if the weather goes sour.

The onus is on the pilot, too. It is our responsibility to maintain VFR and to tell the controller when we can't. Even though National was reporting VFR that day, had one of the VFR airplanes headed toward the airport piped up and told them that he could not maintain VFR on the assigned heading, some measure of chaos could have resulted. It is thus important to make a careful evaluation of weather at a busy airport and go somewhere else if it is a VFR flight and there are any questions about being able to maintain VFR.

Alternate Plan

If going into a busy area VFR and weather conditions are variable, it is much better to head there with an alternate plan toward an area of known good weather. The flexibility to zig and zag to go around showers, for example, may not be there in a congested terminal area. On the trip into National just related, I think we were probably flying along with only one thought after we established ourselves on the Anacostia River: land at National. There were some good divert airfields in the area, though, and had we needed to beat a hasty retreat it could have been done without violating any regulated airspace.

Special VFR

A special VFR clearance is available at some airports, but this is something to be used with a great deal of care. This clearance lowers the VFR minimums to clear of clouds and one-mile visibility. These special minimums are available in the daytime or at night to instrument-rated pilots flying aircraft that meet the requirements for IFR flight. Flying with a low ceiling and a mile visibility is a dicey business and, historically, there have been a number of special VFR accidents. Where it is a useful procedure is on a hazy day, with no real cloud cover and a visibility of, say, two miles. If a pilot elects to fly in such conditions, the advance plan becomes even more important because the navigating will be all the more difficult in that haze. In other words, best have something to follow. The ultimate bad scene is probably in the L.A. basin or the New York area, where a pilot unfamiliar with the area tries to find one of the general aviation airports in a suffocating haze. In both places, the landmarks are hard to locate and the airports are buried in the urban sprawl. Even pilots based at Teterboro have trouble with that airport on hazy days. It can almost be a "stealth" facility.

Haze brings another challenge. Clouds are difficult to see through haze, and when VFR we are supposed to stay out of those clouds. Perhaps the best way of all to navigate VFR to an airport in haze is to have the instrument approach plates on board and just fly the procedure. This would call for being in contact with approach control because any IFR arrivals to the airport would also be flying the approach so traffic would be concentrated there. A night VFR arrival to an airport in a metropolitan area is much the same as a hazy day arrival. Airports can be hard to find at night and a navigational plan is required.

Crosswind

When the wind is really blowing, a crosswind might become a problem. At a controlled airport the tower assigns a runway which is not always the one most directly aligned with the wind. Some towers try to use runways other than the large airplane active runway for general aviation traffic; the result is usually that the airline aircraft get the runway that is most nearly aligned with the wind and we get to remember what those rudder pedals are for. That's okay, if that is what makes the traffic flow. What we need to do is make a plan for the arrival that ensures we are ready to deal with the crosswind as we fly on in. If it sounds like too much, tell the controller and he will work something else out for you. Or you can go elsewhere.

Airport Geography

Another item at big airports is the geography of the place. If we fly to one of these airports, especially at night, the ground navigation after landing can become a substantial problem. The main thing is to listen carefully to all instructions and question anything that is not clear. Ground controllers are alert for airplanes that need progressive taxi instructions, but at a large airport they have a lot of balls to juggle. Even though you might get a clearance to taxi somewhere that is issued without caveats, implying consent to cross all runways between here and there, if you are in doubt, ask.

Hopefully, in the preflight planning for a trip to, say, JFK airport in New York, you studied the airport. They use an inner and outer taxiway system there, the inner being the closer one to the terminal complexes. This has proven to be confusing to pilots unfamiliar with the airport. General aviation parking is off Runway 31L/13R, which makes that the preferred runway for landing in a light airplane, but they seem to prefer the other one, with you then making the long taxi.

If you elect to go to a major airport and don't study the airport geography, much confusion can result—especially after dark when it all may look perfectly logical to an airline captain sitting over 30 feet high in a Boeing but like nothing more than a shimmering sea of blue lights to a pilot whose eye level is 5 feet above the ground. The first time I flew to

Kennedy I got all confused and actually taxied all the way around the circular outer taxiway twice.

Hazards of the Wake

Wingtip vortices from a larger airplane can be a definite hazard at a busy airport, especially in reasonably calm conditions. These are most pronounced when the heavy aircraft is flying slowly, as it would be on approach. There have been problems with a light aircraft landing on a parallel runway with a gentle breeze blowing the vortex over to that runway.

Because the vortices settle after they are formed, the best avoidance procedure is to fly your approach above the flight path of the heavy aircraft and land beyond its touchdown point. This would be true on a parallel runway as well, especially if there is a light wind blowing from the heavy's runway toward yours. In the general traffic pattern area, the heavy will likely be flying a longer final than VFR traffic, so the point of the most likely encounter would be closer to the airport. It is always advisable to watch the flight path of any heavy airplane in the area and avoid the wake. You can be twisted around quite a bit by an aircraft that isn't much larger if you follow closely, especially if the air is calm. Just as wind moves vortices, turbulence tends to destroy them more quickly than they would dissipate in calm conditions.

Another thing to beware is following a heavy airplane that did a go-around. If you visualize the wake turbulence on such an approach, there is simply no way to get to the runway without flying through it. Wake turbulence is one of those things we don't deal with. Avoidance is the only approach, and because it is so potentially serious, always double-check the deliberations and make certain that the action decided upon is correct.

A Blast

Once on the ground at a busy airport, we have to be aware of jet blast. There are not a lot of airports around where general aviation airplanes have to taxi astern of jets that are lighting up or beginning to taxi, but before taxiing behind one, you want to ascertain whether or not the engines are running. If they are, make sure they are at flight idle. Just at idle power, they can rock a light airplane, and some of the larger ones could probably blow you away at flight idle if you taxied close enough behind.

Back to the Country

This discussion was started in Montana, where there was no air traffic, and it evolved to busy airports, which we all think are harder to use than those nice single-strip general aviation airports. Like so much else in flying, though, the simple things can sneak up on us if we don't approach them with concentration equal to that for what seem difficult tasks. It can actually be a greater challenge to arrive safely at an uncontrolled but busy general aviation airport than at a busy one. The challenge here would be enhanced by scuzzy weather, wind, a lot of traffic, or just one other airplane if both pilots' timing happens to be similar.

There is even an interface to consider with air traffic control. This isn't the formal variety of air traffic control, but the informal, the common traffic advisory frequency. This is an important part of our intermingling around uncontrolled airports.

War Story

I'll tell you a tale about this. I happened to be on an IFR arrival that I was making at the time (I'll expand on the IFR technique of this phase of flight in the next chapter). It's the pilot who was trying to be on a VFR arrival who is of interest here.

The weather was far worse than forecast. As I descended toward the minimum descent altitude for the VOR approach to Carroll County Airport in Maryland, my airplane was in and out of clouds and was moving through what might be described as a driving rain. I honestly didn't have a lot of confidence in being able to successfully complete the approach. If I missed, I'd go to Frederick, 28 miles away. The weather there was far better when it was passed a few minutes earlier, and Frederick is equipped with an ILS.

I was having fun, as I always do on an actual approach, and was working appropriately hard to do it all correctly. There is nothing like having a critical peer in the right seat, watching, to dot the "i" and cross the "t."

I could tell early on that I probably would not be able to cancel IFR on the frequency, because I'd be too low to talk to Baltimore who "owns" this airspace. I'd have to cancel on the phone after landing. They had cleared me to the advisory frequency, and I had switched but not yet called in when I sensed that something wasn't quite right.

I wasn't concerned about other traffic because it was clearly IFR and there wouldn't be any other IFR airplanes out there, but what I had sensed was something other than the movement of clouds. Whatever it was caused me to glance to the left and down. Lo and behold, there was an airplane, a high-wing Cessna, one or two hundred feet below, heading in the same direction and apparently trying to dodge clouds.

CTAF to the Rescue

I was almost at the minimum descent altitude by this time, two or three miles from the airport, flying in and out of clouds that the VFR aircraft was trying to dodge. My first reaction was a radio call to "…a Cessna south of Carroll County…" The pilot of the airplane responded promptly. Sounding somewhat harried, he said that he was trying to get into Carroll County but was having trouble staying out of clouds. My thought was that this had to have an immediate resolution, so I told the other pilot he was right below me, I was on an instrument approach, and would he please turn toward the west and try to stay VFR. He shot back with a fast roger. After we landed he called again and solicited some help in getting into Carroll County which, being in Class G airspace, might even have been at legal VFR minimums at the time. The ceiling was about 700 feet, but there was a lot of lower scud. The visibility was about a mile in moderate rain. I told the pilot I wasn't convinced he could make it VFR because of the rain and scud and that the weather at Frederick was much better. He didn't make the paper the next morning, so I assume he made it safely to Frederick.

That experience outlines the importance of using the common traffic advisory frequency (CTAF) whether you think you need it or not. In this case it was used to provide separation between an IFR arrival descending toward Class G airspace and a VFR arrival running the scud. Even though we are all supposed to know that what you see is what you get on weather, the fact that it was worse than forecast that day probably contributed to the other pilot's continuing, thinking he was in a localized shower. It lured another one that day. As we drove home from the airport, a Cherokee flying VFR and apparently attempting a VFR arrival at Carroll County whizzed overhead flying at about 200 feet in the Maryland hill country. Scary, but he didn't make the papers the next day either.

Other Uses

The uses of CTAF aren't always that dramatic, but the frequency designated for that purpose at uncontrolled airports is always useful. Pilots have to back it up with some measure of traffic pattern discipline, which is a key to collision avoidance in a normal VFR arrival. There is also a technique to using the frequency in an uncontrolled traffic pattern. Because it is a party line, state the name of the airport first, to get the attention of other pilots in the same pattern or vicinity. Then you can use your "N" number and state your position and intentions. Some pilots don't use the number and just say, "Carroll County traffic, green Cessna on left base." That probably doesn't conform to the letter of one or more laws, but it is effective.

Standard Traffic Pattern

Some years ago the FAA tried to come up with the definition of a standard traffic pattern. The idea was to define one and then require everybody to fly it. The current rule that all turns in the traffic pattern be made to the left, unless right turns are specified, doesn't define much of anything. And one airport I know of that uses right traffic on one of its runways has probably 10 to 20 percent of the traffic on that runway use a left hand pattern. If they call in and if someone answers, they are told that it is a right pattern. It is also published everywhere. The left-handers in this right pattern illustrate that there's still a need for pilots to make themselves aware of everything about an airport before the arrival starts and for everyone to use the advisory frequency.

The FAA was not successful in defining a good enough standard traffic pattern to consider making it a rule. There is the recommended pattern that is generally taught to students. Once out of the nest, though, most pilots remember that the rule says only that you must make all turns to the left (or to the right if specified for the particular runway), which okays base leg entries and straight-in approaches. How much we are on our own at uncontrolled airports is evident in the Federal Aviation Regulations. The rule telling us about operations at controlled airports, where the tower tells us what to do anyway, is six times longer than the rule covering operations at airports without control towers. That is good—if the rule on traffic patterns had ever been written, the reverse would probably be true. But it clearly signals a "heads up" when we are flying where there is no tower.

How Best?

Because there are only recommendations on standard traffic patterns, the first thing we have to assume for an arrival is that although some might follow the recommended standard pattern, others probably won't.

That means taking nothing for granted about the interface with other aircraft until your airplane is tied down or in the hangar.

A primary rule that I always follow at uncontrolled airports, and one that helps in spotting other traffic, is to be at pattern altitude a few miles away from the airport. Airplanes are easier to see when you are at or below their level. They are hard to spot against a matrix of houses or roads and the paint jobs on some airplanes are pretty good camouflage.

The one pattern entry that I don't like and try to avoid, is the entry on a crosswind leg. The interface with aircraft staying in the pattern, or departing the pattern on downwind leg, is not good when you enter on a crosswind. Climbing airplanes are lower, they are hard to see, and if they don't announce their intentions it is difficult to figure out what they are doing. To me, the better pattern entries are the 45-degree entry to a downwind or the base leg entry. On either of these, traffic in the pattern is likely at pattern altitude and in front of you for better viewing. It has always appeared easier to set up proper spacing when doing it this way. As is always true in the see and be seen system, though, we can't fixate on where we think other airplanes should be. We have to look everywhere, and even those who support the "big sky" theory of collision avoidance, and trust to luck more than anything else, will concede that the closer you get to the end of the runway at an uncontrolled airport, the smaller the sky gets and the higher the likelihood of trying to share the same piece of air with another airplane.

The Final Act
The last part of the VFR arrival at an uncontrolled airport can be very much affected by weather systems—specifically by wind. Many, or most, uncontrolled airports do not have multiple runways, so the chances of having a crosswind are higher. There is usually less runway length, too, so the luxury of landing with no flaps may not be available. It thus becomes a time to sit up straight, take notice, and fly right.

The strongest winds in good VFR conditions usually come in the warm sector ahead of an active cold front or behind a cold front. The strongest prefrontal winds are usually from about 190 to 240 degrees. The strongest postfrontal winds are usually from around 270 to 320 degrees. It depends on a lot of things, but the velocities range from 20 to 40 knots, sometimes higher in gusts. Certainly, strong winds can blow from other directions, but they don't do so as often and when they do, it is usually in connection with some sort of storm. Weather forecasters presume some modicum of intelligence on the part of pilots here, as witness the following, which was part of an area forecast the day Hurricane Hugo came ashore in 1989. "EYE OF HURCN HUGO IS JUST CROSSING THE FIR LINE E OF JAX MOVING NWWD. UNLESS THERE IS AN INCRS IN CNVTN NEAR HUGO THE HURCN SHUD NOT BE A MAJOR PBLM SINCE NO RESPECTABLE AIRCRAFT SHOULD EVEN BE CLOSE TO IT ANYWAY."

Airfield Survey

When planning an arrival on a gusty day, it's best to first do an airfield survey. How many degrees off the runway is the wind? Are there obstructions upwind of the runway that will excite the wind variations beyond the natural gusts? What is the length of the runway when compared with the minimum runway length? (The minimum runway length should be considered to be 160 percent of the book value shown for an approach over a 50 foot obstacle, a landing and a stop. That is what is required in airline operations, and it should provide a good margin as long as the approach speed is correct.) If it is a 90-degree crosswind, all factors have to be taken into account in choosing a direction of landing, including the fact that when wind speed increases in a gust, the wind tends to veer, or shift, in a clockwise direction. This, though, can be affected by upwind obstacles. The margin should be increased to 200 percent of the book number for a 90-degree crosswind to allow for the possibility of some downwind component. Consider the reported wind, but never fail to look at a windsock, preferably at the approach end of the runway. When making a decision on an absolute crosswind, the windsock is the best source of information. After everything is considered, how do you feel about making the landing? If there is much trepidation, best find a runway that is aligned with the wind.

Most airplanes have word on a maximum demonstrated crosswind component. This is not a limitation, it is literally what it says. For example, airplanes that were built in Wichita usually have a higher number than airplanes that were built in Florida because there is generally more wind in Wichita for the demonstrations. Crosswinds, though, don't lend themselves to absolute numbers. Wind shifts and gusts and does all sorts of things. The decision on how much can be handled relates more to what the pilot sees and feels on final approach than to a number you can crunch.

The Landing

The crosswind landing on a runway that is not of excessive length is one of the closer relationships between the pilot and the airplane and the elements. The elements are having a direct and definite effect on the airplane on a second-by-second basis, and they change on a second-by-second basis. Any mechanical approach to such a landing doesn't work. It has to be a fast-paced reaction with the controls moved to what you see and feel. And it is not the smoothness of the touchdown that counts. The primary grading points in this little battle with the elements come from where on the runway you touch, both in relation to the threshold and the centerline, at what speed and attitude you touch, and whether or not the airplane was drifting at touchdown. If you touch down in the first third of the runway, on the centerline, with no drift, and at a normal touchdown speed in a nose-high attitude, you won the battle with the elements. Whether it was a squeaker or a thumper doesn't really matter unless it was such a thumper that the landing gear collapsed or something was bent.

Just because the transition from flight to ground has been made, though, don't slip into that Distinguished Flying Cross just yet. The conflict between the airplane and the wind isn't over until the airplane is secured. Pilots have been known to be so relieved over a satisfactory touchdown that they quit flying and let the wind do unusual things to the airplane during the rollout. Anytime the wind is strong and gusty, the airplane has to be "flown" as it is being taxied.

Technique

What is the best technique for this landing in a strong crosswind on a relatively short and narrow runway? The first item is to relax. A tense pilot is jerky on the controls and will make the ride feel wilder than is really the case. The method used to take out the drift is up to the pilot. I prefer to use the crab method, taking it out in the flare, using the rudder to push the nose of the airplane around and point it down the runway. Aileron is used as necessary to keep it over the centerline. You have to be ready to really push on that rudder pedal, so the nose of the airplane points straight down the runway. You also have to be ready to make the decision that you have the required control power to deal with the crosswind. If the amount of rudder available seems not up to the task, then the elements win and you have to go find a better runway. Anytime you find yourself using the majority of control travel as the flare begins for a crosswind landing, that's not good because as the airplane slows in the landing and rollout process the controls become less effective. It is probably time to go somewhere else. One thing that helps in a crosswind, if there is sufficient runway, is to not use full flaps. On my 210, for example, at 15 knots crosswind I limit flaps to 20 degrees for landing and add five knots to the approach speed. At 20 knots I limit to 10 degrees and add another five knots.

In the Dark

A night VFR arrival is far more demanding than a day VFR arrival, mainly because you can't see as well at night. Most runway lighting systems at relatively small airports are not the greatest in the world, and although some crosswind is acceptable at night I have found that there's no way I can accept as much crosswind at night as by day. It is simply a lot harder to fine-tune the touchdown at night because most landing light systems don't really illuminate the runway as well as that old sun does—even on a cloudy day. Night arrivals have to be better planned than even day arrivals, and approach slope control becomes critical. Some years ago an aircraft flew into trees on final to a relatively small airport nearby. We all wondered how the pilot could possibly have done that. Then I flew there one night and lined up on final, on what appeared to me to be a normal slope. On final I noticed that suddenly the green lights at the end of the runway were no longer visible—a sure sign that I was on a slope that would pass through rather than over high trees off the end of the runway.

VFR arrivals are like everything else in flying. If you go to the trouble to plan them, and to consider all factors, they work much better than if you try to just deal with the situation as it unfolds. When you do them correctly, too, there is a lot of satisfaction—especially if you tack one of those squeaky landings onto the completion of the arrival.

Chapter 11
The IFR Arrival

An IFR arrival is almost an ultimate combination of challenges. We have to do a deal that combines precision flying with weather and air traffic control, one where the finale is the gentle chirp of the tires at the proper spot on the runway. The farther we go into an IFR arrival, the fewer options we tend to have. When we depart or are flying en route, we have a lot of latitude on headings and altitudes as long as the controller approves. The closer we get to the runway, the more we have only two choices: Fly the published approach or go away.

What is the most difficult approach? If I had to pick one, it would be the nonprecision circling approach to an unfamiliar airport. Make no mistake about it, there is a home-field advantage to approaches. Some pilots take this too far, feeling that at their home-field they can do a bit better than minimums. The better deal is to fly the approach to the letter of the procedure, using that home field advantage to keep the risk as low as possible.

The reason the nonprecision circling arrival is hardest is that it combines precise instrument flying and scud running. We do that instrument flying business right up to the minimum descent altitude. There we are turned loose to maneuver visually for a landing. To add to the spice of the situation, runways at airports served by such approaches often aren't overly long, so the challenge of flying the airplane around and getting it established on final at the proper speed and rate of descent is added to the proceedings.

Weather

To raise the ante another buck, these approaches are sometimes flown at airports where there is no weather reporting. This is no factor as long as you follow the prescribed procedure and go away if the airport is not in sight at the proper time, but it is another area where the home-field advantage weighs heavily. For example, when I was based there I could usually listen to the automatic terminal information service broadcast at Dulles and Baltimore and get a fair picture of my chances of a successful approach at Carroll County Airport in Maryland. It wasn't simply a matter of listening, it took some interpolation. The elevation at Carroll is more than 600 feet higher than Baltimore; if there is an easterly or southeasterly flow, it is moving upslope toward Carroll, and the chances are the ceiling will be much lower than reported in Baltimore. For some reason, if it is raining the Carroll weather is usually equal to or better than Baltimore, especially if the wind is from southwest around to north. I don't think that I have ever seen the word used, but perhaps that is a "downslope" condition. Anyway, from the experience of living there I developed some small ability to anticipate actual conditions based on reported weather at other stations.

Carroll County now has an AWOS-3 and the proliferation of AWOS and ASOS means that we have weather reports from a lot more airports.

No Radar

There is no radar coverage at Carroll below about 3,000 feet, so you are pretty much on your own. Certainly it is not a place you would try to go if there were thunderstorms in the area unless you had airborne weather radar and/or a Stormscope or Strike Finder.

Knowing the territory and the landmarks is where the real home-field advantage comes in. Folks kidded me about what I called the K-mart approach, but it sure worked. The eastern edge of the K-mart was aligned with Runway 34 and was about a mile or slightly more from the runway. When you could see the K-mart, if you couldn't see the runway it would be difficult if not impossible to land straight-in on Runway 34. At the other end, on Runway 16, the west edge of the county motor pool lined up with the runway. Having knowledge of these landmarks was helpful, especially on a circle to land on 16 in a high-wing airplane. At an unfamiliar airport you have no such clues to use in fashioning a precise low-altitude circle. Another helpful thing is knowing where the obstructions are that set the minimum descent altitude. At Carroll they are all south of the airport, which doesn't mean that you can bust minimums once by them, but that a circle south of the airport is less desirable than a circle north of the airport. Since I based there, Carroll has gotten a new and longer runway plus GPS approaches with lower minimums. Progress.

Crutches

There are some crutches to use at unfamiliar airports, none of which should be allowed to tempt any deviation from flying an approach exactly as published. But one of the big deals in an approach, especially one where the navigational facility being used is located on the airport, is in having some idea of the distance you have to fly before reaching the airport. This can tell you a lot about whether or not a straight-in approach will be possible. If you are a mile and a half away at 600 feet and can't see the airport, there is a message there. Distance information is also a good clue on when the runway should come into view. Knowing where to look is a big part of finding something visually in restricted visibility. Taking advantage of the information from GPS when flying an NDB or VOR approach can be helpful. The ultimate crutch is a moving map display. The one on the Garmin GNS 530 in my airplane is really neat.

GPS approaches offer another crutch. If there's a GPS approach, there's a waypoint at the end of the runway. That gives the distance left to fly which is a valuable number to have. For example, if you are flying at the MDA and the runway is not visible when a mile out, then you know that the flight visibility really isn't a mile and the approach will be missed.

Not Often

One factor that adds to the challenge of nonprecision approaches is that we don't fly a lot of them, even when we are based at an airport where this is the only approach. And where flying ILS approaches under the hood can have some degree of realism, nonprecision approaches under the hood do not simulate the hard part, from the MDA to the runway, very well. It is here that we have to make good and fast decisions on the circle. On the approach I told you about in the previous chapter (*see* page 123), where I had VFR company while on final approach, the runway came into view at the minimum descent altitude but the K-mart was hidden by a piece of scud. The wind was out of the southwest, so a circle for 16 was desirable anyway. The rain was hard, so although the visibility appeared better than a mile, it wasn't that much better. Later, after the circle and landing, I reflected that it was indeed a challenging arrival. On the subject of that VFR company on final, that is something that has to be considered when arriving IFR at an airport in Class G airspace. A pilot flying VFR has every right to be there as long as the visibility is one mile and he operates clear of clouds. What this suggests to the pilot flying IFR is to start looking for other traffic as you break out. The use of the CTAF frequency is important as well. All that may seem like a lot to add onto the already heavy burden of a circling approach, but the flight isn't over until it is over.

Don't

Some pilots contend that you should not even attempt circling approaches at minimums—especially at night. It is not an easy maneuver, but if it is done properly the risk should not be much higher than on any other approach. If, though, a pilot does not feel comfortable or current at this type of approach, he shouldn't attempt it. And if one is attempted it should be flown with a strong resolve to go to the missed approach at any time in the circle if the visibility drops or clouds get in the way. Just because you saw the runway doesn't necessarily mean that you continue to see it.

However you feel about the circling approach, it is somewhat an anachronism. Even with a zillion bucks worth of avionics in the airplane, this one comes down to the seat of the pants, what we see, and the altimeter. It is a true scud run. The good news is that we have a lot more straight-in approaches with GPS. The circling approach should soon be history except in unusual situations.

Fuel

Because there is a high likelihood of a missed approach on a nonprecision arrival, especially in an area of grungy weather with no reporting at the airport, best do so with plenty of fuel. Flying any approach with a nagging worry over fuel is a bad deal. If I miss this approach, where can I go? is not a question that should have only one answer. Especially if headed toward an airport where you don't have a weather report, best have a foot on another base or three just in case.

The lead up to the approach discussed in the previous chapter (*see* page 123) was a trip in from Champaign, Illinois. The wind forecast that day was totally in error—180 degrees in error to be precise. A closed low aloft had formed and there was actually a strong easterly flow at all altitudes where a westerly flow was forecast. The air traffic controllers were having a hard time with it and were asking pilots for actual winds so that they could update their computers. One pilot reported the wind at Flight Level 220 as being from 130 degrees at 70 knots.

The effect of this on our eastbound flight was substantial. I had loaded plenty of fuel for the trip with the anticipated tailwind, but as we got closer, and the reported weather in the area was much worse than forecast, I started getting antsy about the approach to Carroll County and what might follow if it turned into a missed approach. So I got to make two IFR arrivals—one at Martinsburg, only about 40 miles to the west, and another at Carroll County with the tanks awash with fuel. That made the Carroll approach easier because after a missed approach I would have a whole raft of options. Without the fuel I bought in Martinsburg, the only options would have been nearby Frederick or Baltimore, and landing at either one would have been with minimum legal fuel. The word "hold" would have induced sweating palms.

What If?

The question arises about what you might do if it really gets down to fumes for an approach. Although it is a subject that we should not have to discuss, it is something that occasionally does happen. In the Avianca 707 accident on Long Island, the crew knew they were critically low on fuel and before the approach had discussions on the procedure for a go-around with minimum fuel.

This leads to a hypothetical question: What would you do in a situation where you were flying an approach with minimum fuel, not enough to go anywhere else and an amount that would be questionable for a missed approach and another approach at the airport at hand? The best thing to think about, or go over with the crew, would be how to make the first approach work. After all, a perfectly flown ILS will lead the airplane to the runway, the proper distance down the runway. It might be suggested that not everyone can fly a perfect ILS; on the other hand, virtually all instrument pilots have at one time or another flown a perfect ILS approach, and certainly a condition like this would be the time to pull the best arrow out of your quiver.

ILS

The ILS approach is labeled precision for good reason. It will indeed lead you to the touchdown zone of the runway. As a pilot all you have to do is follow the guidance of the ILS, have the speed and configuration of the airplane at the proper value, and land.

By being a precision approach, though, the ILS does offer some new, different, and interesting challenges. Events, such as wind shear, that may be minor annoyances on a nonprecision approach, become larger deals on an ILS because of its precision nature. Where we may awkwardly recapture the final approach on a VOR approach after descending into a substantially different wind, the same shear may lead to pegging the needles on an ILS and a missed approach. Also, on a nonprecision approach the control of glidepath is not critical up to the point of leveling off at the minimum descent altitude, which is critical. On an ILS we are called on to maintain a precise glidepath while at the same time maintaining a precise track over the ground—all in shifting wind. Earlier I said that a simulated ILS might have more realism than a simulated VOR approach. The exception here deals with wind shear. If you go out and practice ILS approaches in calm conditions, or when there is no pronounced wind shear, you simply won't be ready for an approach where the surface wind is calm and the wind at 1,500 feet is from the southwest at 40 knots, as often is the case.

When?

One thing we can do is charge ourselves up for wind shear on an approach. The first acknowledgment has to be that there will usually be some. We learned in the basic study of meteorology that the wind usually tends to shift clockwise above the level where surface friction is a factor. This in itself defines some wind shear for approaches. Some conditions suggest far stronger shears than would result from this basic state. They would be fronts and thunderstorms. Another basic we learned early on is that fronts have shallow slopes. Once a cold front passes, for example, the surface wind may be out of the northwest and the wind aloft remains out of the southwest for a while. When a front becomes stationary, the surface wind may be out of the east or northeast where a thousand or so feet up the wind is howling out of the southwest. When north of a warm front, the surface wind would be southeasterly and the wind a thousand or so feet up might be out of the southwest. Shear not only affects the heading we have to fly and the power setting that we have to use to stay on the proper glide slope or rate of descent, it also makes for turbulence, which adds a level of difficulty to the approach.

To gain advantage in dealing with shear on an approach, we have to visualize the air that we are going to fly through on the way to the runway and to have some knowledge of how the behavior of this air affects the airplane.

Take, for example, an approach where we know there is a relatively strong southwesterly flow aloft and a light southeasterly flow at the surface. We can easily deduce this from the weather synopsis, from the wind encountered en route, and the reported surface wind. Next suppose that the approach will be an ILS to Runway 6. The shift and change in velocity of the wind won't be much of a factor until we start the descent at the marker. Then we will be dealing on the descent with a decreasing tailwind as well as a wind that is shifting around to come from the right.

The decreasing tailwind will cause the airplane to trend high on the glide slope. Put another way, it will take less power to stay on the glide slope and at the desired airspeed during the descent. It's always a disappointment when someone fails to ask why, so I will assume that you did so. When an airplane flying at a true airspeed of 110 knots has a tailwind of 40 knots, the ground speed is 150 knots. If the wind suddenly goes away, there is no way the airplane can decelerate 40 knots instantly, which it would have to do to have a ground speed of 110 knots. Of course, the tailwind doesn't go away instantly; it goes away in the course of 1,000 to 1,500 feet of descent in two or three minutes. But the airplane still has to effectively decelerate and will thus trend high on the glide slope, or the airspeed will stay too high, if the power setting used is the same as would be used without a decreasing tailwind. If the condition were a decreasing headwind or increasing tailwind, the reverse would be true. The airplane would have to effectively accelerate during the descent to maintain a constant airspeed and the power required would be greater.

Happens

The strongest example I have seen of the decreasing tailwind phenomenon was in my Skyhawk, at Sioux Falls, South Dakota, on an unseasonably warm winter day. The trip from the south had been with a roaring tailwind. The approach was an ILS to Runway 3 and the surface wind was light. I don't recall the exact value, but the tailwind was on the order of 40 knots as I lined up on a long straight-in on the ILS, primed for the decreasing tailwind on final. The strong effect that it had was still a surprise. For much of the approach the power was practically at idle; when I was through the shear, the power had to be brought back up smartly to keep the airplane on the glide slope for the last few hundred feet of descent.

Dealing with the change in wind in relation to localizer tracking somehow seems easier. I knew that day I would go from a direct tailwind, to a crosswind from the right, to very little wind right at the last. I didn't make an anticipatory correction after starting the descent, but was certainly ready to correct the heading to the right as required to track the localizer.

Earlier I mentioned flying with Todd Ericson of the Air Force Academy in an instrument flying contest. We were not grading the contestants on flying the glide slope in this particular contest because it wasn't possible to fly a full ILS with all of them. After telling Todd that there would be no grade on that, I suggested he try a perfect one anyway. He did exactly that, and although there was no strong wind shear there was the challenge of a strong crosswind. When he turned to the inbound localizer bearing without putting in something for wind drift, I thought he had made a mistake. Then, before I could perceive any movement of the localizer needle from my perch, he started a series of mini-turns to the left—a degree or two at a time. After about 10 degrees he stopped. I still hadn't seen the needle move or stop moving and charged this perfect display of technique off to youthful inexperience.

GPS gives us a great tool to use in anticipating and dealing with changing wind as we descend on approach. With the accurate and almost instantly updated readout on track and groundspeed we can get a good idea of the wind that exists as the descent from the final approach fix is begun. This can be compared with the surface wind at the airport to get an idea of how the wind will change as we descend. If there's a strong wind at 2,000 feet and a calm wind at the surface—not unusual in inclement weather—anticipating the change gives the pilot an advantage. The change usually starts at between 1,000 and 1,500 feet above the surface and is usually finished by a few hundred feet above the ground. It can usually be felt by some jiggles in the air and the track readout on the GPS will give instant word when a heading change is needed to maintain the desired track. Track, not heading, is what gets you there.

Worst Case

The worst wind shear on approach would come in a frontal zone or if the approach were flown in air under the influence of a thunderstorm. The two best examples to use in contemplating the latter are airline accidents, the Delta L-1011 at DFW and the Eastern 727 at JFK.

Usually when we think of dealing with both thunderstorms and air traffic control in a terminal area, the feeling is that any problems would come because of less flexibility in being able to zigzag around cells because of relatively high level of traffic in airspace that becomes ever more confined as you move toward the runway. This was not the case in these two accidents. Approaches were being conducted in a normal manner and in both the traffic appeared to be flowing smoothly. In each case the airplane apparently flew into a cell that was on the localizer course as the cell peaked in strength. The L-1011 was in quite heavy turbulence; the 727's ride didn't appear as bad from reading the accident report. But both encountered downdrafts and an increasing tailwind that resulted in the airplane losing airspeed and altitude.

Most airlines have a policy against flying approaches with thunderstorms in the area, but a lot of approaches are flown in close proximity to thunderstorms every year. Passing south of Nashville one stormy late-spring day, with a convective SIGMET calling for tornado activity in the vicinity, I marveled as airline aircraft picked their way around weather and made approaches into Nashville. They closed the airport to arrivals for a while, mainly because the interface between airplanes and storms and airspace had gotten out of hand and there was a raging storm at the airport. Once the airport was clear of the worst, though, they started clearing airplanes in. Looking at the activity, no way I would have flown in there. I just am not interested in a ride that would get that rough.

I'll always remember a story told me by a person who rides airlines a lot and had the wits scared out of him at St. Louis one stormy day. He said that he couldn't believe the pilot would land in such turbulence. On final approach he said it sounded as if the power were going from idle to the stops on a regular basis. He could feel major excursions in pitch and roll, and swore that you could hear the heavy rain beating on the airframe. He saw only ground after the airplane hit, which was how he described a landing that he wasn't sure was successful until he realized the airplane was rolling down the runway. Maybe he exaggerated or maybe the captain really did take a foolish chance.

Quiet Night

Because of the FAA's policy of shutting off departures to airports that are expected to be impacted by substantial thunderstorm activity, we can have some interesting flights right after storms pass by. Such was the case one evening when I was headed to McGuire AFB to talk to the Aero Club there.

A squall line was on the prowl. I had been tracking it all day, and based on the way it was moving I anticipated it would clear the McGuire area between 6:15 and 6:30. So I told them, sure, I would be there in time for the 7 P.M. meeting. When, at the airport, I phoned for the clearance I asked if they could see the back edge of the line on their radar.

The person to whom I was talking was a bit noncommittal but said he thought we'd get to McGuire okay.

The surface wind was howling out of the northwest when we took off in Maryland, but at 5,000 feet the air was smooth. The frequency was quiet because the line hadn't been by Baltimore for long and the pace of inbound traffic had not yet picked up. We could see the back side of the squall line as we flew toward McGuire and could tell that we'd be catching it before we reached the base. My idea of handling thunderstorms and IFR arrivals is to wait until the thunderstorm goes away. The Air Force concurs and McGuire was officially closed to arrivals as we got within about 30 miles. On their ATIS they were advising that there was thunderstorm activity over and within five miles of the airport. It took only about 10 minutes for the activity to move on and we landed there at 6:30. When dealing with thunderstorms, especially in a terminal area, it is amazing what a little patience will do to reduce risk.

Architecture

We have to realize, when flying an approach when there are thunderstorms in the area, that the architecture of the air traffic control system does not allow for a lot of flexibility. On the trip to McGuire the airport was closed and the solution to how to handle our flight was simple — hold. Even had the airspace not been deserted because they were keeping all the airline traffic at bay on the ramp somewhere, it would still have been simple because the altitude and location that I was using for the hold was both away from and under the major traffic flows.

The way terminal areas are designed dictates where we go. They are not all alike, but a typical one has four arrival "gates," 30 or more miles away from the terminal airport, where inbound airplanes enter terminal airspace. Altitudes vary, but low-altitude flights will often go through the gate at 7,000 feet (or lower if they had been cruising lower) and higher altitude flights at 11,000 feet. The center controller may affect the handoff to the approach controller before you reach the gate. If he does, the approach controller won't give you a lower altitude until you get into his airspace. With four arrival gates, they probably also use four departure routes, going between the arrival gates. It makes a nice and orderly way to move aircraft in and out of a terminal area.

The use of radar is a key to managing airspace areas where there is a substantial volume of traffic, but it is used more to monitor, vector, and space traffic than it is used to resolve conflicts. The way the whole system is designed is to keep the controller from having critical conflicts. In other words, when you are arriving IFR, the separation from departing aircraft is procedural. The radar is used to avoid problems more than to solve them.

At the other extreme is the bucolic airport where there is not much traffic and perhaps not even good radar coverage. My base at Carroll County was such a place, and you might think that this resulted in poor IFR service; it did not. After two years and about 800 hours of flying out of the airport, I had not had an arrival delay. Even with a relatively high minimum descent altitude (754 feet AGL), there was only one time when I couldn't get into the airport and that was because the VORTAC was off the air and there was no other approach to the airport at the time. In spite of the greater flexibility at such an airport, you still can't deal with a thunderstorm on the final approach course.

Scorecard

This gets us to the finale on the IFR arrival, a place where we have to be honest with ourselves on the question of human nature and cheating on minimums. We deal with weather and we deal with the air traffic control system. The latter won't do much for us on the former, and where the air traffic control system will catch us if we stray off an assigned altitude, there is nothing there to catch us busting minimums. They may know how high we are from looking at our Mode C readout, but there is no way they can tell what we see. You can cheat all you like without fear of being caught—right up to the point where you fly into fog-shrouded terrain. Gotcha.

There is a strong temptation to cheat, too. I even had an FAA employee once suggest that I go a little lower than the MDA during a nonprecision approach. He knew we were by some radio towers that established the minimum at a relatively high altitude because he had tuned the ADF to the radio station and watched the needle move by the 90-degree position to the right.

A lot of pilots out there firmly believe that the lower you go the more you can see. This is not always true. It depends heavily on the condition that is causing the weather. If it is fog, it goes to the ground, where there may or may not be some visibility; in true ground fog the visibility actually decreases the closer you get to the ground. When an airplane is operated below a minimum descent altitude or decision height and the pilot can't see the runway, the condition of flight can be summed up in one word: perilous.

It might be considered less perilous on an ILS approach, a precision approach, but that does not seem to be the case. In fact, the majority of the airplanes that are lost in minimums-busting accidents are lost on full ILS approaches. It is true that to hit short you have to disregard the glideslope information, but when a pilot feels as if he has to make it in, the slightest hint of being able to see something on the ground may cause him to go for it.

I got a good example of going for it some years ago, when I visited Beech Field in Wichita. Beech was quite a red, white, and blue place under its original owners. When making the RNAV approach to the north, for a straight-in approach to land north, there was a business with a lot of red, white, and blue on its building a mile or so south of the airport. More than once, I hate to admit, as I was flying along at the minimum descent altitude, I started to go for the red, white, and blue of a lumberyard, or whatever it was, only to realize that it was not a building by the runway at Beech Field. This was done with distance to the runway information on the panel, which makes it even more foolish.

The rule is very specific on what you have to see before you can leave the DH or MDA, and it all has to do with approach lights, the runway, and the lights that are related to the runway.

IFR arrivals are, to me, one of the most fun parts of instrument flying. They come with a strong grade, too. If, after an IFR arrival, your palms sweat and you have a bad taste in your mouth, the approach didn't go as well as it should have. Time to practice.

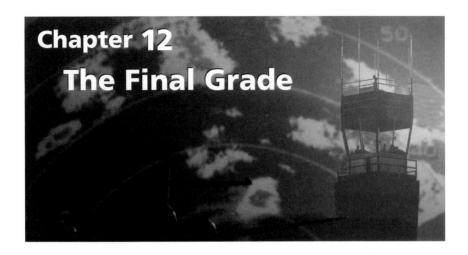

Chapter 12
The Final Grade

When all is said and done, it is how we fit our airplane into the air traffic control system, and how we balance weather flying and risk, that gives us the final grade. It is a rather fine balance that results in value received from the flying—both in satisfaction and transportation—while at the same time avoiding the headlines. The latter is important to all of us. Whenever any one of us has a serious wreck, it affects all.

The accident rate in general aviation flying is not as good as it should be and one of the reasons is that even experienced pilots take foolish risks at times. It is usually done in the interest of being somewhere at a given time, but the tragic results of taking too much risk are felt even by those who didn't know the participants. Higher insurance premiums, less social acceptability, and more regulation are just a few things that stem from the relatively high accident rate. A stranger, not a pilot, cornered me on the subject of safety a while ago. She had some knowledge of a Bonanza accident where, according to what she had read, the airplane had been overloaded. To her, the pilot was an apparent criminal for overloading the airplane. I tried to explain all the factors but lost the battle. In the end, all she would say was, "Tell it to the widows, those guys are dead forever."

The simple fact is that the record is bad. If I were to make a list of all the friends and acquaintances killed in general aviation airplanes, it would take more than one piece of paper. When I said this in a crowd of aviators one day, someone said, "But everyone you know flies airplanes." True, but everyone I know drives cars and rides airliners. The list for those two forms of transportation is one or two names.

So what should you do to keep your name off the list? There are no guarantees because flying your own airplane is almost a total freedom. It is one of the few endeavors left where you can take all the risk you want to take. The challenge is absolute and highly individual. That is the appeal of it to many people. It sure is to me. But you want to manage those risks and not do foolish things. I remember once flying a kit-built aircraft on which an inadvertent opening of the door would have rendered the rather clumsy aerodynamic shape unflyable. Later I chastised myself for flying an airplane that was so poorly designed. The rationale behind the self-inflicted tongue-lashing was that if ever an airplane got the best of me, I would sure want it to be for a more complex reason than the door coming open.

Flying is something that you can do as well or as badly as you like. Not everyone can do it in an easy and relaxed manner—some just lack the coordination between the eyes, the brain, and the hands that it takes to fly well. The way the system works, though, they can try, and, to a certain extent, they can succeed. This raises questions. When you read an accident report of an airliner that crashed when being flown by a pilot who had trouble in training and with checkrides, you have to wonder if the system isn't too lax.

The best way to keep up with how you are doing is to go back through flights and give yourself grades. Be critical of decisions and technique. If you ever thought you heard the angels singing your song, don't let that event go by without giving it much thought. They will remember those words for a long time. If it was that bad, was it worth it?

Examples

I'll use a couple of trips as examples of self-grading flights. The first, from Carroll County in Maryland to Atlanta and Savannah and the return, was flown in a period when there was a stationary front languishing across the route. Whenever we strike out in a frontal zone, the risk blips up a little—how much depends on how dedicated we remain to staying out of convective activity.

I awoke to an uncomfortable sound the morning of the trip south. It was raining so hard out that the sound of the rain was a rather loud roar inside the house. There was no thunder, at least I didn't hear any, but flying conditions are seldom pleasant in rain that hard. The radar reports I procured from my computer suggested that the worst of the rain would move by soon, so I elected to delay the departure for about an hour. What I suspected was happening was that a piece of the stationary front had assumed the characteristics of a warm front and was moving toward the northeast. That is what was making all the noise on the roof.

As we drove to the airport, the weather was improving, with the visibility about three miles in some areas. At the airport I got a bit of a surprise when I called Baltimore Clearance Delivery on the phone to get my IFR clearance. I had filed the route that goes east of Washington over Baltimore, Patuxent, and Richmond, thinking that would be the best way to go. The weather, though, had moved over that route, and good old air traffic control took it upon itself to send me first west to Martinsburg and then southwest toward Atlanta. Mentally, I gave myself a demerit for not figuring that out. It really is the pilot's job to pick the best route. The good news was that the meteorologists that work in the air route traffic control centers were on top of the weather and were suggesting better routings for everyone.

Welcome to the World

All the preflight preparation was okay and the departure was graded as okay. It was what I deem a low visibility takeoff—anytime the visibility is one mile or less I treat it as a low visibility departure—so I transitioned to instruments at liftoff and, as always, was interested in how the airplane reacted as it climbed through various layers of wind. The surface wind was light. I took off on Runway 16 and at first the climb was less than spirited. Sick may be a better word for it. But at about 300 feet the airplane started climbing into a layer were there was a brisk southerly flow and the rate of climb picked up markedly for a while because of the headwind increasing in velocity with height.

Headed out toward Martinsburg, we were in visual conditions at 10,000 feet, but it took only a glance out the windshield to see that this would not last. There was a cloud layer above and a cloud layer below, and it appeared that they would merge up ahead.

After passing Martinsburg and turning more toward the southwest, I could see some rainshowers ahead on the radar. Other aircraft were asking for deviations, and I made the decision that a little to the west of the airway would be the best way to go. This was based on a third hunch, a third what I saw on the radar, and a third what I saw outside. It looked lighter over that way. The center controller wasn't able to offer much other than the fact that he was looking at about 50 miles of rain, but airplanes had been passing through it with no particular complaints.

The Stormscope wasn't showing anything in the direction I was flying, but once in the area of clouds and rain the airplane was bumping through some rapidly building and fairly energetic cumulus. Using the radar on the 10-mile range, I was dodging the heavier showers.

Nonelectronic Decision

After about 30 or 40 miles of rain, I got the visual reward of the morning. About 30 degrees to the right I could see what appeared clear sailing. The controller approved a turn 30 to the right and soon we flew out into lovely weather. The rest of the way to the Epps ramp at Atlanta's DeKalb–Peachtree Airport was in good weather, and on the descent I decided that for the run over to Savannah we might want to go higher than the 7,000 feet I originally filed. Although the cumulus didn't appear to be growing into thundershowers anytime soon, they were growing.

On the ground, after a fairly decent landing, I considered the flight in retrospect and didn't find anything badly wanting. We had landed with plenty of fuel, a couple of hours worth, which some would say is too much. To me, that is open to question. The old saw about the only time you have too much fuel is when you are on fire is one way to look at it. The other way is not to carry the weight of extra fuel. Somehow an extra 90 or 100 pounds doesn't seem like too much to me. On weather, I would have been happier with the flight had I, instead of someone else, made the decision that the westerly route was the best. The quality of the ride through the area of weather after departure was okay and I couldn't think of any way that could have been done better.

Onward

I didn't fly the leg over to Savannah. I picked up another pilot in Atlanta and he flew, with me picking on him from the right seat. But when the guy forgets to pull the chock before starting the engine you do get picking rights. Right? The weather was nice, and although we had to dodge some cumulus tops at 9,000 feet the heading changes were small and the flight smooth. After the landing I considered that the leg had been flown and managed well. The chock was no bother. Anyone who hasn't started up with the chocks in hasn't flown much. The bad technique would be in trying to power the airplane over the chock.

We had four people in the airplane on this leg, and although the weight and center of gravity were well within limits we probably had too much fuel on board. The climb would have been better with 100 pounds less and we still would have landed in Savannah with 150 pounds.

Return

The return flight from Savannah to Atlanta to Maryland was more interesting. The stationary front had sagged southward in the two days we were in Savannah and it had become much more active. There was word of heavy rain and violent weather on the morning TV shows, which is always a tip to watch out. Willard Scott's *Today Show* map had the southeastern United States colored red almost all the way to Maryland. On my weather briefing sheet I wrote of a low-pressure area in south-central Kentucky with a stationary front to the east, across South Carolina, and a trough from the Atlanta area down into the Gulf of Mexico near Pensacola. A thunderstorm was in progress in Atlanta, with ceilings in the area between 600 and 1,000 feet. The radar report was showing thunderstorm activity in the Atlanta area, including some clusters, as well as in eastern North and South Carolina. The only convective SIGMET was for isolated embedded thunderstorms in eastern North Carolina.

It was okay to take off for Atlanta as long as the flying would be done on the basis of it being nice but not necessary to get there. The weather was fine for the departure from Savannah and the Stormscope was doing a good job of defining the activity near Atlanta. From the appearance of the dot patterns, we would be passing just north of the activity. Closer, a controller concurred that we would be north of the worst of the weather, though we would have to fly through a rain area to get to DeKalb–Peachtree. As we neared the area we could see that we would be in rain for a while, but we could see through the rain area and would be below the bases of the clouds. It looked fine to go through. The controller asked what it looked like and when I said it looked okay he offered an airplane without weather avoidance gear a vector through the area behind us. It worked well for both of us. In retrospect, the flight got a good grade. Nothing there that I would have changed. The other pilot was flying and was a little high on the approach, but he fixed that.

On the ground in Atlanta, we consulted the FSS for word on how the thunderstorms looked, and from what they said it appeared okay out over Athens and then toward the northeast.

I had filed for 17,000 feet, but soon after takeoff I started hearing pilots flying above 11,000 complaining of ice. I settled for 9,000 feet, where the airplane was in cloudy but smooth air that was above freezing.

Show Anything?

They probably think it a pain, but when signing on to a new frequency I ask each controller if he shows any weather. That day the answers were all negative until I got up fairly close to Greensboro. Word was that a line of showers started about 40 southwest of Greensboro. The line was about 30 miles wide and the report was that airplanes had been flying through it with no problem. I could see some relatively light showers on the radar and went around those. There was a weak dot pattern on the Stormscope up ahead, and to be on the safe side I asked for a deviation and turned 30 degrees to the left to see what effect that would have on the dots. If they stayed put, it would have been activity far away. If they started to move down on the scope and off to the right rather soon, then it was relatively close. The dots moved. I kept the 30-degree correction until the pattern was at three o'clock and then headed in the desired direction. The deviation was thus a good deal. In a weak frontal condition a few dots on the Stormscope can mean close convective activity. Weak or not, I like to miss it all.

In the vicinity of Richmond we flew into an area of building cumulus. Most of them could be missed visually, but we clipped a few and got some bumps. A Piper Arrow pilot in the area reported strong up- and downdrafts in these cu, but it was really no factor. The weather turned VFR around Baltimore and was VFR for the landing at Carroll County. The landing wasn't bad and there was plenty of fuel left in the tanks. The flight through the frontal zone had been with a due measure of caution, so I felt okay in giving the conduct of the flight a passing grade on risk management. I would take some negative points for initially thinking that it could be flown at 17,000 feet. The signs were there that this wouldn't work, but I didn't read them. A substantial gig came from accepting the short runway for takeoff at DeKalb-Peachtree instead of asking for the longer one. A light crosswind was favoring downwind and the 210 was relatively heavy. Why use a 3,736-foot-long runway under those conditions when a 6,000-foot-long runway was just a short taxi away? You might call that purely foolish.

The interface between the systems, air traffic control and weather, worked fine on the trip. All requests for deviations or altitude changes were granted. Later that day the weather part of the equation went sour. The trough that had extended down into the Gulf turned into an active cold front and there was a major thunderstorm outbreak in the southeast after we were out of the area. My friend Mike Boyd was trying to get back to Maryland from Savannah in his new Mooney, but it just wasn't possible that afternoon. That low in Kentucky had become a major player and the next day, after it moved by, we had record cold in Maryland.

Which brings up my morning theory: When there is a potential area of thunderstorms to deal with, I always feel as if I am ahead of the game if the flight is planned to get through the area between ten in the morning and noon if at all possible, with perhaps a little leeway on either side of those times. In many conditions conducive to storms, they seem to take a break then. There is no suggestion that this always works. There aren't any absolutes on weather. But I have gotten through areas unscathed many times when those trying to fly later in the day have had their troubles, as was the case that day.

More Storms

The next trip was to Little Rock, Arkansas, and back, from Maryland. Patrick Bradley was down visiting that weekend and he was headed back to Teterboro at the same time. His weather appeared to be okay. Our weather to Little Rock appeared awful. The morning news told of tornadoes in Indiana, and when I checked the radar reports there was considerable activity across the route, with one convective SIGMET for a line of thunderstorms. The weather in Maryland was good—some high clouds and unrestricted visibility—so I thought we'd head out and see how far we could get. The flight was in the clear and in smooth air all the way. The storms had moved rapidly south of the route of flight and about all we had to contend with was a headwind that was stronger than forecast. This was one weather system that conveniently moved out of the way. The weather was so good that there was nothing to grade the flight on. I made a couple of nice landings, too, and changed my fuel stop from Nashville to Lexington, Kentucky, to protect fuel reserves from the stronger headwind. Okay on the grade, but no great challenge existed.

Back to the Storms

It didn't work that way for the return. The airplane was getting some new equipment and they said they would finish with it by ten, then twelve, then two, then four. While this was going on, I was watching an area of thunderstorms that started the morning out by beating up on St. Louis and then headed east. This was in connection with a rather weak cold front, but the air was extremely unstable in the area. The storms were so strong at St. Louis that TWA was diverting some if its flights to Little Rock to await better conditions.

Last Check

The last time I checked weather at Little Rock's excellent flight service station it appeared that the thunderstorms were intensifying and moving east faster than anticipated. Where it had earlier been an area of activity, it now was turning into a long line of thunderstorms. Where the earlier prognosis was for the storms to be in western Kentucky by late afternoon, they were going to do far better than that on speed. Still, it appeared that with a slight deviation to the south, it would be possible to go around them on the trip back to Maryland.

Off at 5 P.M. and turned toward Bowling Green, I could see that it was going to take more than a slight detour to the south. The Stormscope was alight in that direction, so I asked the controller if I could swap Bowling Green for Nashville. That was agreeable with him, so I turned right a little to head direct for Nashville.

Faster

They were reading off a new convective SIGMET every hour and I wrote them down to keep up with the movement of the system. It was moving out. The first one I wrote down put the weather 60 west of Nashville, and as I was getting close to there about an hour and a half later, there was no way I was going to fly over Nashville. They shut down approaches while a thunderstorm was in progress at the airport, but then opened back up with thunderstorms still in the immediate vicinity. Traffic was still slightly backed up. I flew about 25 miles south of Nashville, giving the activity a good berth. This line had turned distinctly mean and the convective SIGMET was warning of gusts to 70, two-inch hail, and possible tornadoes. Those aren't fun clouds in which to poke around.

I had originally filed direct to Bowling Green and then on to Charleston and Martinsburg, West Virginia, and on home. After I passed south of Nashville, the controller kept wanting to know when I could head for Charleston. I should have straightened this out for him by filing a new route, but I just hadn't thought about it. I wasn't actually using any of the navigational gear in the airplane, I was using the Stormscope for navigation. Keep the dots five degrees to the left of the nose and fly on. We were at 17,000 feet and a lot of cumulus were building rapidly through that level well east of the line; I was avoiding those visually, but darkness wasn't far away and I wanted to get a bit farther east before heading northeast, in hopes of staying away from and out of those cu. They probably wouldn't have broken anything but would have been extremely uncomfortable.

Wiseguy

When one controller seemed to insist that I tell him when I would be able to go direct to Charleston, I answered "Next Tuesday." What I should have done was define a new route, which I finally did. Direct to Hazard, Kentucky, and then direct to Martinsburg appeared to work. (Somehow "Hazard" does not seem an appropriate name for a VORTAC.) It wasn't risk related, but I deserved a mark for not defining a new route earlier.

By this time it was getting dark and the storms were, if anything, building faster. The line was long, too. The last SIGMET I wrote down started at Erie, Pennsylvania—from there to south of Nashville is a long way.

When it got completely dark, we started seeing a light show ahead and only slightly to the left. The line of storms was moving fast, and when about 120 miles from Martinsburg I asked the controller where the storms we were seeing were in relation to Martinsburg. "About 40 northwest, you'll make it." You can see lightning a long way away at night. If the news was good for us, it wasn't good for westbound air carriers. Each pilot, when he came on the frequency, would ask about ride reports westbound. The controller responded that he had nothing but lousy ride reports to the west.

When I got closer to home, Dulles approach control helped by giving us a turn to the right so we wouldn't have to fly over Martinsburg. That put some extra distance between us and the storms. When I had the airport in sight and signed off with Baltimore Approach, I thought for a moment how well the air traffic control system works. All the pilots that evening were doing whatever was necessary to get the best possible ride and stay out of the dangerous stuff. When that happens, it must increase the controllers' work load enormously. No complaints, though—just the usual excellent service. There was a lot of weather out there, but the controllers were making the system adapt to the needs of the moment. Then I reminded myself that I still had to land on a runway that doesn't look very large at night and to quit thinking about air traffic and pay attention.

G'night Old Paint

As I was putting the airplane in the T-hangar, I had a word with it. Good ride, four and a half hours from Little Rock. Lots of scary stuff out there but no bumps. The only question on grading the flight related to flying halfway across the country starting at five in the afternoon. I told myself before takeoff that if I got tired we'd land and spend the night somewhere. After takeoff I learned that in doing the work behind the panel the technicians had disabled the autopilot, so I would have to hand-fly the airplane home. I don't have much of an autopilot anyway, but the heading hold feature can help to keep the fatigue level down on a long trip. If this were to become a factor, we would land. I didn't get tired, though. In fact, I felt more rested toward the end of the trip than at the beginning. Perhaps this is because the weather got more interesting as the flight progressed. I think that when flying I suffer more immediate fatigue when nothing is going on than when there is a lot of action. The ultimate judge of how tired I am when I fly at night is always the landing. The 210 is not an easy airplane to land well in the dark; that evening I made an acceptable landing, so I didn't feel I had burned too much off the other end of the candle in flying home that night.

I did have to ask myself whether or not the storms had anything to do with the decision to continue. The next morning they might have been right over where we spent the night, or between there and home. Would you rather fly an hour and a half after dark in the late spring in good weather, or wake up the next morning wanting to get on home with thunder and lightning in progress? I don't think I really thought about it that way, but it was probably lurking in the subconscious.

What to Grade?

When looking at flights in retrospect and assigning honest grades to our performance, we have to put greater weight on more important items. The minutiae, though, count too. How well we handle all the chores of communicating and navigating in the air traffic control system has a lot to do with how much undistracted time is available for working with weather information to make the flight as smooth and as risk-free as possible. When we are hand-flying the airplane, any major distraction caused by a glitch in navigation or communication can cause a critical diversion of attention away from the most important task of all—keeping the aircraft under control.

The planning done before flying, and the organization of things as we take off, weighs heavily here. As I think back to flights, the biggest gig that I give myself on a regular basis relates to charts. Properly done, the approach chart for a return and land or the takeoff alternate would always be at hand, and the enroute chart covering the area would be properly unfolded and held in a clipboard. It usually is, but not always. In watching other pilots, I've noticed that some don't have trouble with this but often forget to turn on the transponder. (I leave it on all the time but occasionally take off without setting the assigned code.)

Minor, But...

These seemingly minor items should affect how you feel about a flight when it is completed. Forgetting to unfold a chart or turn on a transponder is a simple omission, one that is often treated lightly by pilot and controller alike. The risk is different, but the action—forgetting something—is the same as might be found when busting an altitude or descending below the minimum descent altitude. Also, if something untoward happened that demanded 100 percent of the thinking and acting time, then such a simple omission could become a major player.

Before we had nationwide radar coverage and Mode C, a pilot error on altitude, clearance interpretation, or VOR setting to track the cleared airway could create a traffic conflict. A classic example of this occurred in the late '60s and should make you appreciate how much better things are now. What may seem a minor gaffe today was much more than that before radar.

A Cessna 310 was inbound to Asheville, North Carolina, where there was no traffic control radar at the time. The flight was cleared by the center controller to the Asheville VOR (since renamed Sugarloaf Mountain), to maintain 7,000 feet. The pilot was told to expect an ILS approach at Asheville. Runway 34 was the only one served by an ILS at that time. Control was transferred to Asheville approach control who said, "...cleared over the VOR to the Broad River; correction—make that the Asheville Radio Beacon...over the VOR to the Asheville Radio Beacon. Maintain 7,000, report passing the VOR." The pilot replied but didn't read the clearance back. Broad River is to the south, on the ILS final to 34; the Asheville beacon is to the north, on final to 16.

The center had told the pilot to expect an ILS approach, so it is logical that had he been planning ahead, he would have had the approach chart out for that. The Asheville Radio Beacon was not shown on the ILS approach chart at that time. What was shown on the ILS chart was a transition from the VOR to Broad River.

When the 310 pilot passed the VOR he called in over it and added, "We're headed for the...for...uh...Asheville now." Asheville approach control simply acknowledged the transmission and cleared the pilot from seven down to six. The 310 pilot turned toward Broad River. In the course of this another aircraft was cleared on the same frequency for an ILS approach to Runway 34 and a 727 was launched on 16 to make a left turn toward the VOR. The 727 was cleared on a different frequency. Almost seven minutes after their original contact with Asheville, the Cessna pilot was cleared for an ADF approach to Runway 16, and to report the Asheville Radio Beacon inbound. The pilot was still flying toward Broad River and acknowledged with a simple "roger."

The departure controller working the 727 thought the 310 was headed for the beacon north of the airport and was basing separation on that. Just over a minute after that last "roger" the 310 and the 727 were involved in what was virtually a head-on collision. Nobody survived.

The value of traffic control radar was underlined by that accident. It simply would not have happened in our present system, but it is a good illustration of how, when not in radar contact, the pilot had best get the navigation right because the people on the ground are basing everything on the pilot following the clearance accurately. If everything is not perfectly clear, we have to ask questions. Today the controller would chastise the pilot, get him turned in the proper direction, and leave it at that.

Major

When grading, any striking out in an incorrect direction would have to be graded as a major error. The best way to keep the airplane on the track anticipated by the system is to double-check everything. When passing a waypoint or VOR, look at the flight plan and check everything. Then check it all again if you are tired or if there is any doubt. If there is a question about the routing, ask the controller.

GPS has done wonders for enroute navigation, and if the flight plan feature is used, which it should be, the GPS will lead you through the flight, automatically switching to the next waypoint at the appropriate time. This and radar shouldn't lull us to sleep, though. Just because you have filed only a few points for a 900-mile trip doesn't mean that the air traffic control system isn't anticipating your track in the same manner it would if you were flying on Victor or Jet airways. If you set up the GPS wrong and it actually has a different waypoint loaded than you filed, things could go badly astray. That is why you need to double-check everything. My Garmin GNS 530 gives you the name of the selected waypoint, and you have to acknowledge this. This can be very helpful to pilots who grew up with one set of identifiers only to have them changed. The most confusing ones to me are Charleston, West Virginia, which changed from CRW to HVQ; Dayton, Ohio, which changed from DAY to

DQN; and Wilmington, Delaware, which changed from ILG (with an interim change that didn't last long) to DQO. For some reason I can remember that IND changed to VHP because wonderful Roscoe Turner, winner of many air races, who was an FBO in Indianapolis, was one of the world's all-time best Very Hot Pilots.

Serious Business

Although the relatively minor things have to be important, the major ones become critical items in grading a flight. Often it is in how we handle an in-flight change in plan that determines whether the grade is good or bad. For example, if the weather at the destination is below minimums despite the forecast of better weather, a diversion to a suitable airport to await better conditions would call for a good grade. Holding for forecast improvement and using up all the reserve fuel would call for a bad grade. If could even get the ultimate bad grade if fuel exhaustion resulted.

How we deal with enroute weather is a big part of the grade because, as we have seen for years, many general aviation aircraft are lost en route because of weather. VFR or IFR, the grade comes when the weather conditions are not good for the flight at hand. For example, an IFR pilot who flies without weather avoidance and depends on vectors "through the lightest part of the line" really doesn't get good grades. The lightest part might be more than can be handled. The pilot who takes a vector to avoid the whole area of weather, or who diverts to wait for better conditions, gets the best grade. A pilot who starts VFR in good conditions but is flying toward poor conditions has to be graded on the quality of the alternate plan, and the use of that or another plan when the original idea turns out to be a bad one.

Approaches

Approaches, especially ones to minimums, are the easiest to grade. On one not too long ago I failed to adequately compensate for wind and blew well through the approach leg on the procedure turn. I got it all put back together, but the grade was still not good. On another, a circling approach to minimums, I got it just right. When I rolled out on final approach right on the VASI and with everything else in place, it was satisfying. Got an A+ on that one. On another, I have given a lot of thought because there is a strong lesson. Also a circling approach, there was a tailwind on base leg and I just didn't adequately compensate for it. The result was a steeper turn than I like down low — 30-degree banks are okay in the pattern and are considered in setting reference speeds for approaches in jets, but I prefer 20. Still, the airplane was not quite aligned on final at the completion of the turn and was high on the VASI. Again, it was relatively easy to put a patch on all this and land, but that didn't make me feel any better about the conduct of the approach. All that would have been required to make it better would have been a full consideration of the effect of the wind on the ground track as I maneuvered for landing.

The really bad grade on an approach would come with any control problems or any violation of minimums. Those things, if continued, will eventually lead to bad things. Some have suggested that pilots are like cats and have nine lives. The trick is in knowing how many have been used up. Personally, I used eight before I was 21.

It doesn't have anything to do with the air traffic control system but the landing is affected by weather systems. It is also where our trusting and faithful passengers sit in judgment of pilots, unless we really do something strange during the rest of the flight.

I was picking up my wife, Ann, at Baltimore one day and hoped the flight would be on time and the ride pleasant, with more emphasis on the latter than the former. There were a lot of thunderstorms out and I hoped the crew would do their best to stay out of them. I watched the Delta jet bringing her from Atlanta land and could tell from the window in the terminal that it was one thudding arrival on the runway. When Ann came through the jetway I asked her how the flight was. "Terrible landing."

There is nothing that we can do about a grade like that but grin and bear it and take silent satisfaction about making it smoothly and safely through the minefields of thunderstorms, ice, below VFR conditions, wind, the air traffic control system, and the rest to get the airplane to the point where the remaining grade is on the touchdown and taxi to the chocks. Bliss would be getting an A+ on every landing, but some things are simply not attainable.

Index

E

Elkins, WV   2
emergency   99, 109
Ericson, Todd   112
experience, flying in weather   11

F

"FAA approved"   9
FAA inspectors   9
familiarity with system   100
Federal Aviation Regulations   7, 67
feeder fixes   92
FL230   107
FL290   107
flight following   49
flight service specialist   25
Flight Watch   89
flying
   direct   45
   expense   11
   risk   7, 63, 109
   routes   104
   smooth   77
fog   24, 57, 80
fog, shallow   57
forecast, long-range   14
four-corner system   43
freezing rain   20
freezing-level forecasts   86
frequencies, sector   113
frequency 278   ix
frontal zone   19, 74, 81, 85,
   136, 142
fronts
   cold   10, 18, 19, 68, 85, 134
   occluded   18
   stationary   18, 19, 21, 68, 134
   warm   18, 20, 68
fuel   8, 132
   minimum   88
   minimum legal   133

G

gate, departure or arrival   43, 138
general aviation airplanes   63, 83
geography   104, 120
GPS   45, 71, 136
GPS approach   131
Grand Canyon collision   x
Grand Prairie, TX   14
groundspeed   136

H

Hagerstown, MD   97
handoff   45, 46, 138
haze   119
headsets   60, 98
high-lift devices   4
holding   111, 138
homebuilts   6
Hudson River   94
Hurricane Hugo   126

I

icing   7, 58, 79, 86
   airframe   30, 79
icing gear   30
IFR
   arrival   129
   clearance   94
   flight plan   37, 41, 47, 77, 94
   release   41
   reservation   99
   routes   93
ILS approach   131, 133, 134
in-trail spacing   42
inflight advisories   82
inflight visibility   72
inspection, annual   8
instrument approach   43
instrument approach charts   103
instrument rating   63
insurance companies   6

W

wake turbulence   *39, 121*
Washington Center   *44*
Washington National Airport   *117*
weather   *107*
  big picture   *81*
  briefings   *27, 77*
  computer model   *14*
  depiction chart   *66*
  en route   *31*
  forecasting   *14, 24*
  marginal   *63*
  minimums   *8*
  reporting   *130*
  synopsis   *135*

weather conditions
  stable   *77*
  unstable   *77*
Wichita, KS   *29*
wind   *73, 117*
  changes in   *2*
  checks   *117*
  velocity   *88*
wind and temperatures aloft   *13*
wind shear   *17, 51, 80, 83, 134*
wind shear turbulence   *46*
winds aloft   *5*
winds aloft forecast   *56*
windsock   *15*
wingtip vortices   *121*